The essential guide to retail lighting

Achieving effective and energy-efficient lighting

Cosmin C Ticleanu, Paul J Littlefair and Gareth J Howlett

The research and writing for this publication has been funded by BRE Trust, the largest UK charity dedicated specifically to research and education in the built environment. BRE Trust uses the profits made by its trading companies to fund new research and education programmes that advance knowledge, innovation and communication for public benefit.

BRE Trust is a company limited by guarantee, registered in England and Wales (no. 3282856) and registered as a charity in England (no. 1092193) and in Scotland (no. SC039320).

Registered office: Bucknalls Lane, Garston, Watford, Herts WD25 9XX
Tel: +44 (0) 1923 664743
Email: secretary@bretrust.co.uk
www.bretrust.org.uk

IHS (NYSE: IHS) is the leading source of information, insight and analytics in critical areas that shape today's business landscape. Businesses and governments in more than 165 countries around the globe rely on the comprehensive content, expert independent analysis and flexible delivery methods of IHS to make high-impact decisions and develop strategies with speed and confidence. IHS is the exclusive publisher of BRE Trust publications.

IHS Global Ltd is a private limited company registered in England and Wales (no. 00788737).

Registered office: Willoughby Road, Bracknell, Berkshire RG12 8FB.
www.ihs.com

BRE Trust publications are available from
www.brebookshop.com
or
IHS BRE Press
Willoughby Road
Bracknell
Berkshire RG12 8FB
Tel: +44 (0) 1344 328038
Fax: +44 (0) 1344 328005
Email: brepress@ihs.com

The Publisher
IHS BRE Press
Garston
Watford
Herts WD25 9XX
Tel: +44 (0) 1923 664761
Email: brepress@ihs.com

Printed using FSC® or PEFC material from sustainable forests.

FB 56

First published 2013

ISBN 978-1-84806-322-8

MIX
Paper from responsible sources
FSC® C016379

The authors are grateful to iGuzzini for permission to use the photograph on the front cover of the Napapijri shop in Milan, Italy.

Contents

BRE will help to make your retail business more profitable, safe, sustainable and energy efficient.

We can help with:

Building solutions

- Use our Independent expertise for technical reviews of standards and specifications for new construction, refurbishment, fit-out and management of your buildings.
- HVAC and lighting design and specification
- Fabric and façade treatment
- Interior fit-out materials performance and sustainability
- Controls, energy management and staff procedures
- Problem solving, including testing and expert witness services.

Corporate social responsibility (CSR)

Boost your brand value and profitability with including:

- Bespoke sustainability policies, targets and implementation plans
- Sustainable retail property frameworks, design guides and information for suppliers
- CSR assessment and reporting.

Safety

Protect people and property from fire and security breaches with:

- Our expertise on building security and fire safety standards, design and operation
- Approved security, passive and active fire safety, and fire detection and suppression, products, systems and services – as listed in the LPCB Red Book – www.redbooklive.com

Sustainability

Our world-leading expertise in built environment sustainability includes:

- Energy efficiency
- We will help you to plan, implement, monitor and continually improve energy efficiency measures and strategies that will reduce your costs and carbon footprint.

BREEAM

Developed by BRE, BREEAM is the international sustainability standard for buildings that is applied to new and existing buildings, refurbishments and fit-outs throughout the retail sector. It is used by developers, investors, clients and occupiers to improve their buildings and the retail environment.

www.breeam.org

Sustainability benchmarking

BRE runs the International Sustainability Alliance which benchmarks the sustainability performance of building assets, and is used extensively in the retail sector in Europe and around the world.

Retail sector clients include:

- Asda
- Audi
- British Land
- Co-operative Group
- Corio
- Cushman & Wakefield
- ECE
- H&M
- International Council of Shopping Centres (ICSC)
- John Lewis Partnership
- Marks & Spencer
- Morrisons
- Redevco
- Waitrose.

www.bre.co.uk

Foreword

The John Lewis Partnership welcomes this new publication from the BRE Trust. It comes at a critical time with energy costs increasingly affecting a store's profitability, and with the price of electricity predicted to increase dramatically, there is increasing interest among retailers in the creation of effective, energy efficient lighting schemes.

Reducing our impact on the environment is a key issue for The John Lewis partnership. Our 2013 Responsible Development Framework sets out our aspiration to design, operate and manage our estate so that it meets the needs of our customers, Partners and the wider community, and has minimal impact on the environment during construction, operation and disposal.

A key feature of the Responsible Development Framework is our desire to focus on trialling new technologies and looking for new ways of working. Successful methods are then applied across our estate.

Against the backdrop of a growing business, we are working to reduce absolute operational CO_2 equivalent emissions by 15% by 2020/21 against a 2010/11 baseline.

Energy is the single largest contributor to our operational carbon footprint. Energy is also a fundamental requirement for us to run our business and rising electricity, gas and oil prices have an impact on the cost of running our business. Developing new solutions that reduce energy consumption, cost and carbon are key to driving sustainability and profits.

Lighting is critically important to retail environments and plays a lead role in reducing energy consumption across all our properties.

The objectives for our retail lighting schemes include attracting customers into our stores, honestly displaying merchandise to its best advantage and to provide a positive customer experience and ambience which aids their navigation and journey through our stores. Factors that need to be taken into consideration when developing a scheme include reliability, maintenance, flexibility, and capital, installation and running costs from energy requirements.

We have been monitoring and testing the technical developments in LEDs for many years but previously found they were not viable for our sales floor lighting due to insufficient output and colour rendering. The availability, affordability and performance of LED lights has advanced at a remarkable pace in recent years due to the multitude of different solutions helping create a greater demand for better technology.

Following successful trials at Waitrose Stratford upon Avon and John Lewis 'At home' Ipswich stores we now specify LED lighting as standard throughout the Waitrose and John Lewis estate for new and refurbished stores. We have proven that LEDs provide significant savings in the electricity required for lighting in Waitrose and John Lewis. We have also completed a successful retrofit of LED lighting which were installed throughout Waitrose Cirencester. This gives us the confidence to roll back this technology across our estate.

Overall, we estimate LEDs will deliver at least a 10% reduction in store electricity demand compared to traditional lighting solutions. It is a great example of how we have worked with our supply chain to pioneer and drive industry to develop new lighting solutions that reduce energy consumption, cost and carbon.

For other retailers looking to develop a low energy lighting strategy we recommend BRE Trust's *The essential guide to retail lighting*. It provides retailers, architects and designers with detailed practical information and guidelines to explain how best to develop energy efficient, cost-effective lighting solutions for new and existing stores.

Nigel Keen

Director of Property Services

John Lewis Partnership

© iGuzzini: Desigual, Capucine, Paris, France

Preface

This guide was produced with the support of BRE Trust as part of its research programme. The survey of retail lighting was originally carried out for the Defra Market Transformation Programme, and the data are reproduced with permission.

The authors would like to thank Feride Sener, who carried out a review of international research on retail lighting, and produced some of the diagrams and photographs.

We would also like to thank Bill Wright of Wright Energy and Environment, who gave a presentation to a retail lighting workshop held at BRE, and to all the participants at the workshop who gave expert views on the subject. Photographs and case studies are printed with permission from retailers, lighting designers and manufacturers as acknowledged within the guide. We would like to thank them for their contributions.

© iGuzzini: Marttiini, Helsinki, Finland

Glossary

Accent lighting

Highlighting of displayed merchandise or the features of a store or building.

Average illuminance

The illuminance averaged over a specific area. In practice, this may be derived either from an average of the illuminances at a representative number of points on the surface, or from the total luminous flux falling on the surface divided by the total area of the surface. Measured in lux (lx).

Ballast or control gear

Part of the control equipment of fluorescent or discharge lamps, used to stabilise the current. The older, traditional mains-frequency ballast can consume up to 20–25% of the total lamp current. A modern electronic ballast working at high frequency consumes about 30% less current, and can be used to regulate or dim the lamp output.

Colour appearance

Visual sensation correlated with the 'warmth' or 'coolness' of the light emitted by a lamp. The metric used to characterise the colour appearance of the light emitted by a light source is the correlated colour temperature, expressed in Kelvin (K). Warm white light is produced by lamps having a colour temperature below 3000 K (reddish hues), whereas 4000 K and above (bluish) is cool and cold white light. The colour appearance is quite separate from the colour rendering of the lamp.

Colour gamut

A measure of how colourful a scene will appear under a given light source. Sources with a high colour gamut generally make objects appear brighter and more colourful.

Colour rendering

The ability of a light source to show surface colours as they should be, usually in comparison with a tungsten or daylight source. Measured on the colour rendering index (CRI) scale. A value of 0 means it is impossible to discern colours at all, and a score of 100 means no colour distortion. For most indoor lighting applications a value of at least 80 is recommended.

Compact fluorescent lamp (CFL)

Type of fluorescent lamp, designed to replace incandescent lamps. CFLs use less power, and have a longer rated life.

Correlated colour temperature (CCT)

See colour appearance.

Dichroic reflector

Lamp reflector used for display lighting that allows heat to pass through while reflecting the light in a cool beam that does not heat the display.

Directional lighting

Lighting that uses lamps or luminaires with reflectors that direct and control the light so that the light on the area or object of interest comes predominantly from a particular direction.

Discharge lamp

Lamp that produces light by discharging an electric current through a gas (neon, argon, krypton) or a gas/metal vapour mixture (mercury, sodium). For some types of discharge lamp, such as fluorescent tubes, the electrical discharge generates ultraviolet radiation, which is converted into visible light by a phosphor coating on the inner face of the glass. The type of coating determines the spectrum of the light emitted.

Driver

Device that converts supply power into the current and voltage that the semiconductor material contained in LEDs requires to emit light. The driver may also sense and correct for changes in intensity (dimming) and colour (tuning) during operation.

Efficacy

A measure of the effectiveness of a lighting installation in converting electrical power to light. Units are lumens per Watt (lm/W).

Emergency lighting

Lighting provided as a separate system for use when the main lighting fails.

Enhanced capital allowances (ECAs)

ECAs are available for the installation of energy-efficient equipment, including lighting.

Extra-low-voltage lamps

Small tungsten halogen lamps producing two or three times the light output of conventional filament lamps. Normally powered from a separate 12 V source. Apart from increased efficiency and lamp life, the heating effect is lower, which is preferable for display lighting.

Fibre optic lighting

Type of luminaire that uses a fibre optic cable to transmit the light from a lamp to a point of use. The lamp is therefore remote from the application, allowing ease of access for maintenance etc.

Flicker

Visible oscillation in the luminous flux.

General lighting

Substantially uniform lighting of an area without provision for special local requirements.

Glare

Discomfort experienced when lighting is excessively bright when viewed against the surroundings. Often the result of inadequately shielded lamps.

Illuminance

The level of light on a surface; measured in lux (lx). Previously called the illumination value. Can be used as a reference measurement of the performance of a lighting system as related to the activity. (See lux and maintained illuminance.)

Incandescent lamp

Lamp in which light is produced by a filament heated to very high temperature by an electric current.

Induction lamp

Electrodeless lamp in which an induced electromagnetic field, generated by induction coils, excites the mercury atoms in the glass tube, causing them to emit ultraviolet (UV) radiation, which is converted to visible light by the phosphor coating on the inside of the tube.

Indirect lighting

Lighting used to illuminate an area by reflection from other surfaces. The fraction of the emitted luminous flux reaching the working plane directly, assumed to be unbounded, is 0–10%.

Initial light output

The luminous flux from a lamp after 100 hours of operation.

Installed load

See power density.

Kilowatt-hour

The standard measurement of electrical consumption; equivalent to one kilowatt of electrical demand for one hour (kWh). Also known as a 'unit' of electricity.

Lamp efficacy

The ratio of the light output from a light source to the power consumed; measured in lumens per Watt (lm/W). The higher the efficacy value of a lamp or lighting system, the more energy-efficient it is. For example, the efficacy of a 60 W incandescent light bulb 12 lm/W, and of an 11 W CFL is 55 lm/W. For a 36 W fluorescent tube it is 91 lm/W. This last figure is a lamp efficacy; it excludes the power consumed by the ballast that is needed to run the discharge lamps.

Lamp lumen maintenance factor (LLMF)

Ratio of the luminous flux emitted by the lamp at a given time in its life to the initial luminous flux. Measures how the output of a lamp decreases with time.

Lamp survival factor (LSF)

The fraction of the total number of lamps that continue to operate at a given time under defined conditions and switching frequency.

LED lamp

Lamp containing a semiconductor LED that converts electrical energy into light and a driver that converts supply power into the current and voltage required by the semiconductor material to emit light.

Light output ratio (LOR)

Basic measure of the efficiency of a luminaire; the ratio of the light emitted by the luminaire to the light output of the lamps contained within it. Depends on the quality of the materials used, as well as on the basic design of the luminaire.

Light pollution

Many older luminaires serving roads and car parks lose light skywards, causing night-time sky glow. Apart from wasting energy, the appreciation of the night sky is lost and astronomical observation is more difficult. Luminaires with a cut-off have a much reduced effect on the night sky. Light pollution can also take the form of unwanted spill light to neighbouring properties and gardens.

Louvre

Open grid attached to a luminaire to control the light output, the prime purpose being to prevent the lamp being seen from normal viewing angles.

Lumen (lm)

SI unit of luminous flux, describing the quantity of light emitted by a lamp or received at a surface.

Luminaire

Light fixture or fitting; an electrical device used to create artificial light and/or illumination, by use of an electric lamp.

Luminaire efficacy

Light output of an entire luminaire (light fitting) divided by the total power consumed by the lamps and ballasts. Equal to the lamp efficacy multiplied by the light output ratio of the luminaire. Units are lumens per Watt (lm/W).

Luminous flux

Amount of light produced by a lamp; usually measured in lumens.

Lux (lx)

SI unit of illuminance or amount of light on a surface. One lux is equal to one lumen per square metre (lm/m^2).

Maintained illuminance

Illuminance at the time when maintenance is expected to take place. Most values of illuminance that are quoted as applicable to a store, eg 500–1000 lx, refer to this value. The time taken to reach the maintained illuminance level would depend on the lamp types and application.

Normalised power density

Power needed per m^2 of floor to achieve an illuminance of 100 lx on a horizontal plane with general lighting. Units are Watts per m^2 (W/m^2) per 100 lx.

Power density

Installed power of a lighting installation per unit area. Units are Watts per m^2 (W/m^2).

Power factor

A measure of the phase difference between the voltage and current in an alternating current (AC) supply. For lamp types other than incandescent, the voltage and current waveforms are not exactly in phase with one another: thus the volts multiplied by the amperes in the circuit may be higher than the Watts. In such cases, the Watts represent the active power, and the volts multiplied by the amperes represent the apparent power. The power factor is the ratio of the absolute value of the active power to the apparent power. Ideally, the power factor should be as close to unity as possible. A low value of the power factor increases the current load and the energy consumption. Most high-wattage lamp circuits are designed to have a power factor greater than 0.85.

Restrike time

The time period between switching a lamp on and its achieving full brightness from a 'hot start', ie the lamp has just been switched off.

Run-up time

The time period between switching on a lamp and its achieving full brightness from a cold start. The time can vary between seconds for a tungsten filament lamp to several minutes for a mercury vapour lamp.

Service period

The average lifetime of tungsten lamps, or the period by which replacement is due for fluorescent or discharge lamps.

Tubular fluorescent lamp

Fluorescent lamp in a tubular form. Light is produced when electricity excites mercury vapour, which in turn produces short-wave ultraviolet light that causes a phosphor to fluoresce, producing visible light.

Tuneable LED

LED lamp capable of varying the colour of the emitted light.

Tungsten filament lamp

Incandescent lamp with a tungsten filament that emits light when it is heated.

Tungsten halogen lamp

Incandescent lamp with a tungsten filament contained within an inert gas and a small amount of halogen. The tungsten and halogen cause a chemical reaction that increases the lifetime of the lamp, and prevents the darkening seen in tungsten filament lamps.

Watt (W)

Measure of the electrical power that a light source consumes. Part of the power input is transformed into light (visible radiation), and the rest is considered as loss (heat and electricity). For example, incandescent lamps transform 95% of the electric power input into heat and only 5% into light, whereas fluorescent lamps and LEDs typically transform 80% of the electric power input into light, with 20% lost as heat and ballast losses.

© iGuzzini: Simulation based on Colonys Shop, Zurich, Switzerland

1 Introduction

This guide reviews typical current practice in retail lighting, and gives advice on how to achieve effective and energy-efficient solutions, providing retailers, architects and designers with practical guidelines both on refurbishing existing lighting systems and on developing new lighting solutions for their stores.

A checklist is included in the Appendix at the end of the guide to provide quick guidance when carrying out refurbishment of existing lighting schemes, particularly when the existing installation is more than 10–15 years old.

Lighting is critically important in retail environments. Light attracts customers, creates atmosphere, and defines image. Good lighting makes a store inviting from the outside and creates the right atmosphere for the selling process, showing the products at their best and stimulating impulse buying. It also improves the way the space is used and the time that people spend in the store, and provides good working conditions for the staff.

However, the retail sector consumes more lighting energy than any other commercial sector. Based on data published by DECC[1] in July 2012, in 2011 retail lighting consumed 35% of the energy consumption for lighting in the entire service sector, and 43% of the electricity used in the retail sector.

Although lighting products are becoming more efficient, longer opening hours and higher light levels have increased the energy used for retail lighting. Also, inappropriate control strategies and improper choice of light sources result in energy being wasted.

There is substantial potential for energy saving and carbon reduction. Reducing the energy consumed by retail lighting can directly increase profitability and reduce environmental impacts. Modern lighting techniques and equipment, and more efficient light sources, provide opportunities for significant reductions in the use of energy, while achieving a greatly enhanced level of illumination and improved visual appeal[2]. Cutting wasted energy for lighting can reduce overheating in a store, and therefore cut the high cost of air conditioning.

Quality should not be neglected when implementing measures to increase lighting efficiency, and therefore attention should always be given both to the effectiveness and the quantity and quality of lighting in retail spaces. For best results in specific retail spaces, a lighting professional should be consulted, ideally in the early stages of project development.

© iGuzzini: Stella McCartney, Rome, Italy

2 Design principles

2.1 Functions of retail lighting

Lighting is an essential component of retail spaces. It has three primary objectives[3]:

- To attract the customer: Display windows and store interiors are both important for attracting customers, and the quantity, quality and effect of light can be used to create a suitable image and appearance that is attractive to the target audience.
- To initiate the purchase: Lighting should guide customers around the store and display the merchandise to the best advantage, stimulating the buying process. Buying decisions start when the customer is visually intrigued, and are usually completed when the customer has visually evaluated the merchandise.
- To complete the sale: Adequate lighting at the point of sale is vital for completion of the transaction, when both sales staff and customers need to read prices and manipulate the means of payment. Lighting different from that in the rest of the store can be used to indicate the location of the payment area.

Research has also demonstrated lighting's role in evoking an atmosphere. A field study involving 57 clothing stores in the Netherlands[4] revealed the effects of different lighting attributes (brightness, contrast, glare and sparkle) on the image of the retail space, and on shoppers' perceptions, feelings and moods. The findings showed that sparkle adds most to space liveliness, whereas the brighter the space, the more unfriendly and less relaxing the atmosphere.

Hence lighting must be carefully addressed to evoke the correct atmosphere within the retail space. It can also provide subsequent benefits by enabling customers to remember the atmosphere and image of the store after the transaction has taken place, and hence can encourage them to revisit it.

Additional lighting systems are required to provide security after closing, and to facilitate egress in case of emergency[5].

2.2 Factors to be considered in retail lighting design

To meet the objectives of retail lighting, several factors need to be taken into consideration:

- appearance of the space, people and objects
- branding
- visual comfort, health and well-being
- energy and running costs
- reliability and maintenance
- flexibility
- capital and installation costs.

The ways in which these aspects are approached differ according to the type of retail space, and the type of merchandise. The main categories, with typical lighting solutions, are as follows.

- Supermarkets or hypermarkets: High levels of general lighting provided by fluorescent lighting, although increasingly LED ambient lighting is being used. Supplementary accent lighting of key areas, using metal halide, compact fluorescent, or LED light sources.
- Retail 'sheds': Typically these are on out-of-town trading estates. They are likely to have high levels of efficient fluorescent or, often, high-bay metal halide lighting. In some cases, though, this has been supplemented by 'mezzanining' – inserting an extra floor or part floor, with lower ceilings and a return to a more conventionally lit space in retail lighting terms.
- Clothing retailers (Figure 1): They may have lower levels of general lighting, with accent lighting on displays. The dominant source varies from store to store, either compact fluorescent or metal halide, supplemented by other forms of lighting, including LED and concealed fluorescent. As a rough rule, the national chains are likely to opt for a more energy-efficient mix of sources, whereas some smaller, independent retailers may still have inefficient tungsten halogen spotlighting as the sole light source.
- Department stores: Typically these have efficient general lighting (usually fluorescent), with a greater amount of accent lighting than in supermarkets or retail sheds. There may be a wider variety of accent lighting to help delineate different areas of the store. In-store franchises are now common; these may have their own, built-in lighting.
- Financial and other services retailers: These include banks, building societies, estate agents, insurance sellers and travel agents. The traditional lighting approach has been to mimic office lighting, often using recessed fluorescent louvred downlighting. More recently, though, there has been a trend towards adding display lighting, often in compact metal halide or LED form.
- Smaller general stores: These include corner shops, hardware stores, chemists, small electrical retailers and charity shops: Usually fluorescent lighting is in use, although it may be old and relatively inefficient compared with current best practice.
- Smaller, boutique-type shops: These include hairdressers, beauty parlours and jewellers: Tungsten halogen and compact metal halide tend to be the dominant light sources in such outlets.

Figure 1: Clothes stores: lower levels of general lighting with accent lighting on displays (© iGuzzini)

- Car showrooms (Figure 2): These have high levels of daylight, but often also high levels of electric lighting with compact sources (usually compact metal halide) to add sparkle.

There are various retail premises that do not fit easily into the above categories: they include bookshops, garden centres, funeral parlours, CD and music stores, art galleries, and mobile phone shops.

Cafés and restaurants have not been included in this guide, but many of the same principles will apply.

Research has shown that the same store environment can be perceived differently under different lighting settings meant to evoke particular atmospheres. A study in Belgium[6] simulated the lighting settings of three supermarket chains targeting different market segments: high-end, mid-level and hard-discounter. Lighting parameters (illuminance, colour temperature, balance of general lighting and accent lighting) were set to replicate the actual conditions found in real stores within the three chains. Findings showed that the store character can be created through lighting, and that the way shoppers perceived its atmosphere was not affected by their mood before entering the store. The high-end lighting settings made shoppers perceive the retail space as livelier, cosier and less tense.

The lighting scheme can vary according to the store profile, from high-level uniform general lighting without accent or display lighting in low-budget stores, to low levels of general lighting combined with intense accent and display lighting in exclusive stores. General lighting and some accent lighting are both typically used in value-for-money and quality stores.

Figure 2: Car showrooms: a combination of high levels of daylight and electric lighting (photo taken at Audi, Tamworth)

Figure 3: High bay lighting used in a DIY store (© David Wright and licensed for reuse under the Creative Commons Licence)

The lighting scheme must also be designed to suit the nature of the merchandise, which dictates the properties of the lighting sources and equipment required to emphasise particular characteristics, such as freshness of food products, or warmth and cosiness for home products.

In addition, lighting requirements differ as the functional requirements of the space change. There are basically four types of space that define a retail environment: sales areas, display windows, ancillary spaces (fitting rooms, workrooms, wrapping and packing, stockrooms, restrooms, lounges, offices), and exterior spaces. Each of these areas has its own functional requirements, which consequently result in different lighting solutions being used[3].

2.2.1 Appearance of the space, people and objects

In most cases the first point of attraction is the display window and the approach to the store, which provide a link between potential customers and the products sold in the store. Highlighting the merchandise by making it appear brighter than the ambient conditions has a significant impact upon potential customers.

Within the store, the size and shape of the space, as well as surface reflectance, colours, and textures, all contribute to the overall appearance.

There has been a trend in recent years towards suspended ceilings, particularly split-level ceilings that allow concealed lighting. Lighting of walls makes the retail space appear welcoming, not gloomy. Another trend is towards the use of coloured lighting for special effects, particularly by the use of LEDs.

In some cases, drama can be more important than a warm welcome. Shops designed to attract young people, such as computer games stores, may have very dark interiors, but with dramatic accent lighting on selected products.

In other cases there is insufficient emphasis on specific lighting of displays; instead there is a bland overall ambient level of light. Well-designed display lighting can reduce the overall installed load by putting the light where it will have maximum impact. Cleverly designed, efficient shelf-lighting solutions allow good lighting of merchandise while reducing illuminance in circulation areas. This strategy increases the contrast ratio between products and their surroundings, and hence makes the merchandise more attractive.

Correct modelling of people is of particular importance in changing rooms and other areas, such as retail premises that sell cosmetic products.

The appearance of the light fittings themselves is also important, as it is part of the overall ambience. At one end of the market (Figure 3), high-bay fittings may be chosen in a retail shed to convey the appearance of a warehouse, making customers think they are purchasing a bargain. At the other end (Figure 4), tiny recessed or concealed 'invisible' lighting conveys an atmosphere of sophistication, with clear, uncluttered walls and ceilings.

Figure 4: Atmosphere of sophistication created by the use of spotlights that create 'drama' effects (© iGuzzini)

Projector and floodlight-type lamps are perceived as adding drama, because they are used in theatres.

Prismatic controllers or wrap-round diffusers on fluorescent luminaires are suitable for retail spaces from an illumination point of view, because they give good lighting on vertical surfaces. These fittings look old-fashioned, though, and are now rarely used. Retailers may instead opt for office-type recessed louvre fittings, as they look more modern and upmarket, even though they tend to light vertical surfaces less well, and give shadows on walls and ceilings. Louvres and lenses on fittings are seen as being more up to date.

2.2.2 Branding

An important aspect of the lit appearance of the store is the way in which it conveys the retailer's brand. This covers issues such as:

- type of retailer (a jeweller's shop, for example, has a specific lit ambience)
- position in the market (upmarket/downmarket)
- type of customer, particularly their age group
- colour schemes for stores
- the way particular signage is coloured and lit.

Lighting is very much a part of the branding dimension of retail marketing. Research has shown that varying the lighting of retail spaces can convey a very different image in terms of brand identity[7]. Lighting can be used to enhance brand communication, and to define the image of a retailer more clearly, generating added value for the business. In upmarket boutique shops (Figure 5) the lighting will be sophisticated, with concealed fittings and highlighted displays. In the 'budget' end of the retail market the lighting and shopfittings should look 'cheap and glaring', so that the customer believes that the products on sale are the cheapest available.

Figure 5: Concealed fittings for display lighting (© LED Light Sheet)

CASE STUDY Waitrose, Bracknell

Waitrose Bracknell continues the John Lewis Partnership's tradition of producing well lit, customer-focused retail environments while driving down energy consumed within the store.

Solutions in place at Waitrose Bracknell contributed to an 8 W/m^2 reduction in energy consumption for lighting compared with older stores (as communicated by Waitrose). Lighting schemes are designed around the store layout so that light is only focused on product. This has been achieved by installing a combination of T5 fluorescent general lighting with accent lighting using metal halide (MH) lamps, and LED fittings in fridges, cold rooms and on illuminated signs; light pipes for daylighting that are integrated with the artificial lighting by a daylight-based dimming control strategy; automatic presence control lighting in the warehouse, staff areas and on the shop floor after trading hours[8].

The design focus was centred on driving down lighting levels through efficient application of technology while ensuring the customer experience was enhanced. The reductions achieved delivered this experience as well as a 30% reduction in energy consumed by lighting (as communicated by Waitrose).

Figure 6: (a) A combination of T5 fluorescent general lighting with metal halide (MH) accent lighting and LED lighting in fridges. **(b)** Integrated fittings combining daylight from light pipes and electric lamps are used above cashiers (© Waitrose)

For these reasons, chain retailers usually use a common design approach in all their stores. This can have negative consequences for energy consumption, because it can make such retailers unwilling to change to more efficient lighting solutions. In some cases branding may conflict with what constitutes a sensible lighting solution. For example, computer game and music stores are often lit like nightclubs to attract teenagers, even though this makes it impossible to read the small print on the merchandise.

However, a common design approach can have benefits too. If a standard design is to be adopted, retailers will usually spend more effort developing it, and may give greater consideration to the energy and maintenance cost consequences of lighting all their stores in this way.

2.2.3 Visual comfort, health and well-being

As well as making the retail space appealing to customers, lighting systems also provide adequate visual conditions for sales activities to be carried out efficiently and comfortably. Therefore lighting must conform to human needs, which are extremely complex: emotions, moods, actions, perceptions and health are all influenced by light.

We all need good light to see well, but light has a much more far-reaching meaning for human beings. It also plays a vital role in maintaining the physiological and psychological balance, through its influence on hormone secretions, body temperature, cognitive activities and moods. These are determined mainly by the level of illuminance, the spectral composition of the light, the light colour and the dynamics of the light source. Correct lighting ensures a subjective sense of well-being for the users of the lit space.

All these aspects are particularly important for staff who work in the store, as they spend a significant amount of time in their workplace.

Visual comfort is an essential human need that is strongly related to task performance, health and safety, and mood and atmosphere. An incorrectly designed lighting system, or a disregard of design specifications, usually leads to visual discomfort. This can give rise to an extensive list of symptoms for the users of the space: red, itchy, sore and watering eyes, or even headaches and migraines.

Aspects of visual comfort include the avoidance of glare, a balanced luminance distribution on the working plane and within the entire visual field, and the colour balance of the indoor luminous environment. Illuminance levels that are too low or too high have a negative impact on visual comfort, improper light output distribution or unwanted light direction may lead to glare, and light colour also affects visual comfort, owing to its psychological effects on people.

2.2.4 Energy and running costs

The retailer's running costs will include lighting maintenance (see Section 2.2.5) and energy costs.

Energy can form a significant cost for retailers. Despite the rise in electricity prices, though, energy costs are small compared to the daily income a store generates: 0.5% to 1% of turnover is typical. Apart from high energy consumption, intense use of electric lighting leads in turn to extra energy being required for cooling. The high heat gains to a space can result in many stores being uncomfortably hot, even in cool weather and with the air conditioning running.

Low power factors (see Glossary) can also increase energy consumption, but this is a secondary effect, because they increase losses in the supply cabling. (Circuits with a low power factor have currents that are not in phase with the voltage, and the current is higher for the same wattage.) If a retailer is billed according to kVA (the product of voltage and current), there is an incentive for them to choose lighting with a good power factor. Most commercial lighting contains corrective capacitors to set the power factor close to unity.

2.2.5 Reliability and maintenance

There are many individual sources in retail lighting schemes, many with a relatively short life. This means that lighting maintenance is a major problem in retail environments, and it is common to see blown lamps, or the wrong lamps, in fittings.

Lighting maintenance is particularly important to retailers, for the following reasons:

- It affects the appearance of the store.
- It is a significant running cost. It comprises not just the cost of the lamps, but the cost of employing someone (either in-house or a contractor) to replace them.
- Maintenance can interfere with retailing, particularly in stores that are open 24 hours a day.
- Maintenance is not simply a matter of replacing lamps. Over time the output of a lighting installation will decrease, as a result of four factors:
  - Lamp lumen maintenance: The output of discharge and LED lamps decreases with age, even if they stay clean. New types of lamp are now available with improved lumen maintenance.
  - Lamp survival: As the installation ages, some lamps will fail. Unless they are replaced immediately (spot replacement), the output of the installation will drop. Ballast failure, if it occurs, will also require maintenance action.
  - Light output ratio: As the luminaires get dirty, output will drop.
  - Room surface maintenance: Some of the light in a room is reflected from the ceiling and walls. As these surfaces get dirty, the level of reflected light will be reduced.

The normal way to allow for these losses is to oversize the lighting system. As the installation ages, the output drops until – when re-lamping and cleaning are due – the system should be delivering exactly the recommended amount of lighting. Figure 7 shows how this approach works.

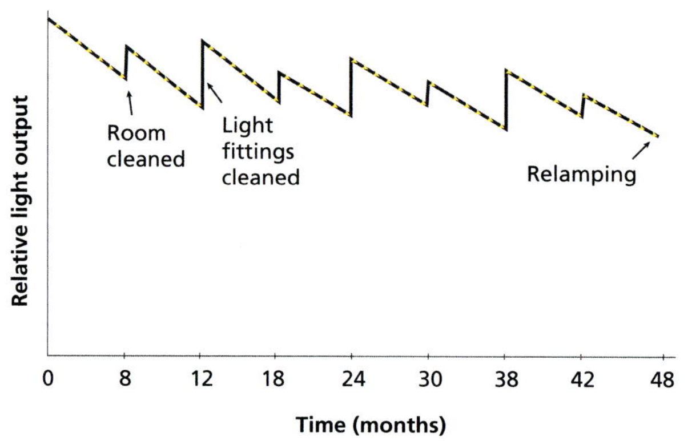

Figure 7: Changes in illumination with time and maintenance. The graph assumes that the room surfaces are cleaned every six months, and the light fittings once a year. Re-lamping occurs after 48 months.

Thus ease of maintenance is not simply a function of lamp life. It also depends on lamp lumen maintenance, whether there is a noticeable colour change, and on how noticeable a lamp failure is.

For these reasons, fluorescent lamps are generally viewed as easy to maintain. The lamps last a long time, and the new tubes now available give very good lumen maintenance. They are often installed in multiple lamp fittings (as many as six lamps per fitting), so if one lamp goes it can often remain unnoticed for some time.

LED lamps generally have a longer lifetime, but at the end of their life the entire fitting often needs to be replaced. LED lamps can be used as direct replacements for less efficient alternatives such as tungsten halogen lamps, if the fittings are compatible, in which case an option could be to insert LED replacement lamps when lamps need replacing. If replacing low-voltage tungsten halogen lamps, it is necessary to check whether the LEDs will be compatible with the low-voltage transformer in the circuit. Also, some LED lamps may overheat and perform poorly when installed in downlights in an insulated ceiling.

High-bay lamps, in contrast, are more difficult to maintain. Although the rated life of the lamp is not much shorter, they suffer worse lumen depreciation, and if a lamp blows, its absence stands out much more.

Tungsten halogen lamps are expensive to maintain because of their relatively short life, and because a large number of individual lamps are required to give a reasonable light level.

Induction lamps are fluorescent lamps that operate at radio frequencies. Because they have no electrodes, their life is very long – comparable to that of LEDs – and they are particularly useful for situations where maintenance is difficult. They may last around 60,000 hours, and have good luminous efficacy, similar to that of a CFL. However, they have yet to be widely used in retail applications because of their high capital cost, although they may be used in hard-to-reach places such as the top of atria.

In general, the most efficient lamps also tend to be those that require least maintenance, whereas the less efficient ones require more. This is another incentive for the retailer to install energy-efficient lighting.

There are two main lighting maintenance strategies. In a spot re-lamping strategy, lamps are replaced as they fail. In a bulk replacement strategy, all the lamps are replaced together at the end of a maintenance interval, which may be two years for fluorescent lamps. Under this latter strategy some spot re-lamping may also be needed between maintenance intervals, to replace rogue lamps that have failed early.

Spot re-lamping has several disadvantages. Maintenance is much more frequent, and at any one time there are likely to be more faulty lamps, either not working or just about to fail. There is more scope for replacing lamps with others of the wrong colour. In addition, lamps that are still lit may be giving a fraction of their rated output, reducing the overall illuminance in the store. This has energy consequences, because retailers may install extra lighting to compensate for it.

Bulk replacement uses more lamps, but often proves to be the cheapest strategy for the retailer, because the bulk of the maintenance is done at one time. For this reason most major retailers use bulk replacement.

Many retailers hand their lighting maintenance over to specialist contracting firms. A typical service might involve bulk replacement of lamps, plus spot replacement if a specified number of lamps fail in the store. The contracting firm will also clean the fittings, which is particularly important if light levels (and hence relative energy efficiency) are to be maintained. Typically they might clean the fittings at the time of bulk replacement, and perform an interim clean halfway through the maintenance interval.

2.2.6 Flexibility

Flexibility allows retailers to rearrange displays and change store layout without carrying out major alterations to the lighting systems.

There are two main ways to provide flexibility. One is to provide a high level of general lighting with closely spaced fittings, so that the entire store is lit without shadows. However, this approach has repercussions for energy consumption.

The other approach is to use adjustable fittings that can be modified when the layout changes. This often implies the use of smaller lamps, usually CDM, although square fluorescent or compact fluorescent fittings the same size as ceiling tiles can be used to good effect in department stores if 'plug and play' wiring solutions are used. The fitting can be moved to a new position easily in place of another ceiling tile.

Track-mounted lighting gives the most flexibility, but a more recent trend has been towards multidirectional, gimbal-type fittings that can swivel within a trunking (Figure 8). They give a neater look to the ceiling. Such fittings usually feature CDM lamps. Another development is a dual-function fitting that can either be recessed in or brought out of the ceiling.

Lighting controls can provide extra flexibility via scene setting, or by providing the ability to switch off certain lighting as displays are moved around.

2.2.7 Capital and installation costs

Capital cost is particularly important to retailers. Typically, a store refit or a new store will be given a set capital budget. Lighting costs have to come out of this budget, along with all the other costs; so if more expensive lighting is chosen, something else may have to be sacrificed to compensate for this. The fact that efficient lighting may save on running costs at a later date is often not taken into account, because running costs are usually part of a separate budget. However, running costs are

Figure 8: Multidirectional gimbal fitting that can swivel within its trunking (© Concord Retail Lighting Fixtures from Havells Sylvania Europe)

important; a typical (80 W) luminaire may cost £100 to £200, but use more than £500 worth of electricity over a 15-year life. Installing more expensive, but more efficient, luminaires can be highly cost-effective.

Enhanced capital allowances (ECAs) are available for the installation of energy-efficient equipment, including lighting. The ECA allows 100% of the capital sum to be written off against corporation tax in the first year after installation, thereby offsetting the initial capital cost of the higher-quality lighting scheme. Detailed criteria have been published covering high-efficiency lighting units (luminaires), white LED luminaires and lighting controls[9]. Equipment has to meet all the criteria to be eligible for ECA support. Manufacturers should be able to advise whether their products comply.

It is particularly important for retailers, and for their shopfitters and electrical contractors, that lighting is quick and easy to install. A store generates large amounts of money from the day it opens, but none at all beforehand. So opening the store on schedule, or even ahead of schedule, is a major driver.

This increases the drive towards standard lighting designs. Such a design reduces both installation time, because contractors are more likely to be familiar with the products, and design time. This can make retailers less willing to change to more efficient lighting solutions.

Some lamps are easier to install than others. For example, standard high-bay luminaires are viewed as very quick to install. In contrast, some CDM metal halide lamps are viewed as difficult to install, as the lamp base design means it is possible to break the lamp when installing it.

2.3 Recommendations for retail lighting

Retail environments vary according to the type of merchandise or services on sale, the location of the outlet, and the target customer group. There are many types of store, and an effective lighting scheme has to be designed to meet the needs of the store in the most appropriate fashion[2]. The lighting system should allow easy, accurate examination of the characteristics and qualities of the merchandise, and minimise glare and unpleasant brightness contrasts.

The three main components of a retail lighting scheme are:

- light levels
- contrast and highlighting
- colour properties.

2.3.1 Light levels

Good retail lighting essentially consists of finding the right balance between general lighting, accent lighting and display lighting, depending on the store profile[5]. Consequently, the levels of illuminance for general lighting to be used vary with the store profile, and can range from low levels (100–200 lx) to high levels (500–1000 lx), depending on the store image.

Low-budget stores, normally used for mass merchandising, are based on a general lighting, single-system approach, with no accent or display lighting. Average illuminance for general lighting levels are high, typically in the range 500–1000 lx, directed onto the merchandise (ie on the vertical faces of the shelves in the case of a supermarket).

High-end, exclusive stores employ complex lighting systems, with a widespread use of accent, display, decorative and ambient lighting. In such cases, a lower average illuminance for general lighting, typically in the range 100–200 lx, is necessary for the accent and display lighting to be effective. A zoned approach to regulating illuminance levels within the space can also create contrasts[3].

Shops that combine value for money and quality profiles typically use intermediate lighting schemes, with some accent lighting for speciality areas and an average illuminance for general lighting on the merchandise in the range 250–500 lx. Department stores often fall into this category.

The general illuminance level in a store has a substantial influence on its overall energy consumption. Although current practice is based on the above criteria, depending on the type of store there are opportunities to decrease illuminance levels, and therefore the lighting energy use. The BS EN 12464-1:2011 recommendations[10] for retail premises differentiate between sales areas, where the recommended average illuminance level is 300 lx, and till areas and wrapper tables, where the recommended average illuminance level is 500 lx. Nonetheless, BS EN 12464-1:2011 also remarks that the recommended values for sales areas are determined by the type of store[10].

Reducing the general illuminance in a store has a direct impact on energy consumption, as the power consumed is roughly proportional to illuminance. It also has secondary benefits:

- Accent lighting has more impact, so it may be possible to reduce the amount of (and power consumption of) accent lighting. High general illuminances can 'wash out' any lighting used to highlight particular displays, so that a high-wattage lamp is required to give accent lighting enough 'punch'.
- Energy consumption and running costs for cooling are lower. This is particularly beneficial for stores in indoor shopping centres, where all the heat generated by the lamps may have to be removed mechanically. In other store types it may be possible to eliminate cooling altogether.
- Daylight (where it is provided) contributes more to the ambience of the store, and it may be possible to make use of daylight and dim or switch off some of the lighting. If a very high illuminance is specified, daylight may not be able to meet the lighting needs of the space for most of the year.

Illuminance is a major concern to retailers for other reasons as well. Some retailers are aware of their main competitors' approaches, and of the ranges of illuminances provided. They may be anxious to ensure that their stores do not look gloomy compared with others.

A high internal illuminance makes a store look open when viewed from the street. This is particularly important in perimeter areas, where the lighting may be competing with the high daylight levels outside. For this reason, store managers sometimes complain about photoelectric (daylight linked) control of lighting in perimeter areas. A compromise approach could be to leave lighting switched on when it can be seen from typical sight lines into the store. Alternatively, signage can be used to indicate that the store is open.

Illuminance calculations are generally carried out for an empty space. In a retail environment the overall illuminance can drop as light is absorbed by stock such as dark clothes, or items in brown or coloured boxes. So retailers will tend to opt for a high illuminance to allow for this.

The Disability Discrimination Act has had an impact on retailers. Those with low levels of general lighting may consider increasing them to help partially sighted customers and wheelchair users.

A sensible approach would be, in principle, for retailers to choose different basic illuminances depending on the location of the store. For example, a higher illuminance would be appropriate if a store were accessed from the open street, and if it were close to other competitors. A lower illuminance would be better if the store were accessed from a relatively less well-lit mall, or if there were no competition nearby. In practice, though, retailers tend not to vary their general illuminances, having a standard value for all their stores.

2.3.2 Contrast and highlighting

Most stores fall within one of two well-defined lighting categories[2]:

- functional, eg supermarkets or DIY warehouses, which require an overall light level.
- theatrical, eg fashion stores, jewellers and high street stores, where lighting is used to draw shoppers' attention to the merchandise on display.

Regardless of the store profile, general lighting should be uniform[5], with an illuminance uniformity (minimum/average) of at least 0.7. Accent lighting can then be used to provide contrast to attract the attention of the customer. In this way the merchandise can be lit more brightly than the overall store ambience, and lower light levels of background lighting can be used to save energy.

The effectiveness of accent lighting is determined by the extent to which the luminance of the merchandise lit is higher than the luminance of its immediate background. Different luminance ratios give different intensities of highlights and shadows[5].

Stores of the functional type usually have a constant light level throughout the sales area, with little contrast between zones. In a theatrical-type store, which requires an element of presentation, a relatively low level of background lighting is used, plus a range of techniques to enhance the space by creating contrasts and highlighting, such as[2]:

- spotlighting, which adds 'sparkle' and interest to display the merchandise to best effects, and also provides a degree of flexibility for subsequent adjustment and repositioning of displays and merchandise
- creating lower light levels for circulation paths than for display areas, to trace out the path of the walkway
- focal points of display lighting used to attract customers and draw them onwards through the store
- perimeter lighting to help to create the image of space, which can be crucial to a customer's impression of the store
- 'wall-washing' techniques to add interest to the space, enhancing textured wall surfaces and wall displays.

Significant energy savings may be obtained in a retail space by changing the lighting design philosophy

Significant energy savings may be obtained in a retail space by changing the lighting design philosophy. Instead of aiming for an overall level of illuminance (and indirectly the level power consumed), this approach concentrates on the lit effect of the scheme, and hence on delivering lighting to where it is required. There are two main techniques:

- Task/ambient lighting (Figure 9): Low illuminance levels in the bulk of the store, but with accent lighting on displays, to highlight them and draw the shopper to key focal areas. There is a trend towards miniaturisation, which results in localised lighting fitted to individual displays rather than the building fabric. Although the individual light sources are not particularly efficient, because they are more likely to be focused on the display they may give a lower overall wattage than an installation in which larger sources are aimed more generally.

  As shown in Figure 9, interior architects and airport retail specialists The Design Solution created this dramatic retail environment, enhanced by carefully integrated localised lighting details within the wall bays and directional highlighting from the ceiling. By taking this compositional approach to the lighting design, the need for ambient light was eliminated.

- Lighting the perimeter of the space (Figure 10): Studies[11] have shown that people tend to judge the brightness of a space by looking at the walls. By making the walls brighter, the whole space looks better lit. Linear fluorescent is often used to good effect in lighting the edges of retail spaces, either recessed into the ceiling, or integrated into the tops of wall displays. This gives valuable emphasis to these displays, and makes a store as a whole look bigger and brighter. One chain of retail shed-type stores used to have double the illuminance (from high-bay lamps) around the perimeter for this reason, but then opted instead for concealed fluorescent wall-washing, thus saving energy and producing a better-lit appearance.

Figure 9: Lower ambient lighting, combined with accent lighting on displays, draws shoppers to key areas (© The Design Solution)

Another possibility is to use daylight to provide some or most of the ambient light. Daylight gives excellent colour rendering and a natural feel to the retail environment. US studies[12] have shown that retail sales rose in stores where skylights had been fitted compared with those where they had not. Daylight can save significant amounts of energy used for retail lighting, provided a suitable control system is fitted. See Section 4.4.

2.3.3 Colour properties

There are two aspects of colour that need to be taken into consideration in retail lighting: the colour appearance of the light sources, and colour rendering.

Colour appearance describes whether the light from the lamps looks 'warm' or 'cool', and makes a major contribution to the message the lighting and the store send to potential customers. A cool, light appearance tends to convey a business-like ambience, whereas a warm colour appearance provides a homely feel. As a general rule, the colour appearance of the light sources used changes from cool to warm as the store profile moves from low budget to exclusive[5].

The colour appearance can also be determined by the nature of the merchandise: for example, chiller cabinets look fresher and white goods look crisper and cleaner under a cool light, whereas gold jewellery looks more attractive under a warm light.

Daylight access in the store requires the correct choice of colour appearance that blends well with it. In daylit interiors, intermediate-colour lamps will blend best with the natural light.

The correlated colour temperature (CCT) of a light source (see Glossary) determines whether it looks 'warm' or 'cool'. This is given in the manufacturer's literature (Table 1).

Another aspect of colour appearance is colour consistency between different lamps on the shop floor. When replacing fluorescent or discharge lamps in an installation, it is essential to use lamps of the same colour temperature, and preferably from the same manufacturer. Some discharge lamps may shift in colour over time; ceramic metal halide lamps tend to be more stable than their quartz counterparts.

The colour of LEDs can vary significantly from chip to chip as they are manufactured. They are then sorted into 'bins' – sets of chips with a similar colour output. The 'width' of the bin (the maximum colour difference between different LEDs in the same bin) is a key issue. Specifying chips from a narrow bin width will give greater uniformity of colour, but this advantage needs to be offset against the higher cost of a tight bin specification.

Colour rendering is the ability of a light source to reproduce object colours realistically. Good colour rendering is essential in helping to generate sales, particularly in fashion retail, where

Figure 10: Perimeter lighting improving space appearance and size (© acdc)

Table 1: Colour temperatures of conventional light sources

Light source	Colour temperature (K)
Daylight	6500
Fluorescent (daylight)	6000–6500
Fluorescent (cool white)	4000
Fluorescent (white)	3500
Fluorescent (warm white)	2700–3000
Compact metal halide	3000–4400
LED	2700–6500
Tungsten halogen	2800

the customer must be able to see the actual colour of the goods on sale. Good colour rendering is also critical in food retailing, where the goods must look appealing.

The light sources should be carefully selected to give the most appropriate colour rendering for the application. Colour gamut is a measure of how colourful a scene will appear under a given light source. Sources with a high colour gamut generally make objects appear brighter and more colourful. Future lamps for display lighting, where it is important for objects to look attractive, could be specially engineered to have a high colour gamut.

However, while it is always important to display merchandise to best effect, it is an offence to mislead customers about the quality of the goods on sale[2].

BS EN 12464-1:2011 also recommends a CIE (International Commission on Illumination) general CRI of at least 80 for retail premises[10]. Varying the light sources used in different spaces (eg sales area and fitting rooms) may change the colour appearance of the merchandise in different parts of the store. Although the CIE general CRI is a useful guide, the final choice of light source is best made by viewing the lit objects of interest[5], which can be of real assistance in judging the colour rendering properties for specific merchandise colours.

© Philips Lighting: Co-operative Food, The Strand, London, England

3 Review of current retail lighting provision

3.1 Energy consumption for retail lighting

Substantial amounts of energy are used for lighting in the retail sector. According to the data published by DECC[1], in 2011 the retail sector used 20% of the total energy consumption in the entire services sector. In the same year the retail sector consumed more lighting energy than any other commercial sector. Lighting energy consumed in retail amounted to 13.7 TWh (terawatt-hours; a terawatt is a billion kilowatts). This represented 35% of the final energy consumption for lighting in the entire services sector, and 43% of the overall electricity consumption in the retail sector[1]. These are average values, as lighting energy consumption depends on the type of store (Figure 11).

There has been an increase in energy consumption for retail lighting, partly because of longer opening hours, and partly because retailers' central driving factor is their need for higher sales. There has been a focus on putting merchandise in a brighter light than in adjacent stores to make it sell better. However, a new factor has made its presence felt more strongly recently – energy price. Predictions by both government and various industry forums show that demand for energy is expected to exceed supply by 2015, with immediate effects on the price of energy, which is estimated to increase by over 20%, particularly in the commercial sector[13]. Increases in energy price can act as a driver towards efficiency, and encourage energy-efficient retail lighting solutions.

The most straightforward way to improve energy efficiency is to reduce light levels. Some retailers have already started implementing this measure in pilot projects, but the overall tendency is to wait either until competitor stores implement this measure first, or until the lighting industry starts helping them change the design and concepts of retail lighting. Most retailers still base their selling strategy on higher general lighting levels, rather than on attracting shoppers by contrasts and highlights on the merchandise.

In the UK there is currently no applicable legislation limiting installed loads and energy use in stores. In contrast, the US state of California, which is one of the strictest places in the world for controlling power consumption, limits lighting loads in stores[14] to 15 W/m^2.

Limits on energy use can also be set by shopping centres for their tenants. The owner of a new shopping centre in Johannesburg, South Africa, set a limit of 34 W/m^2 for lighting energy consumption. Oreb, a designer clothing store in this shopping centre, installed LED fittings that helped achieve 18.9 W/m^2, which was 84% less than Oreb's other stores, where lighting loads[15] vary from 115 to 150 W/m^2. The LED fittings had a colour temperature of 4000 K, and good colour rendering – between 80 and 98 CRI. Given the reduced energy consumed for lighting, the payback period is less than 2 years.

3.2 BRE survey on retail lighting

BRE undertook a survey of retail lighting in early 2011. The aim was to improve the current understanding of lamps in use across the retail sector. The strategy was to survey stores from a representative town (Watford), extrapolating the data to estimate the UK-wide distribution. Just 'front of house' lighting was surveyed. The survey was split between four groups of retail premises: shopping centres, high streets, retail parks and supermarkets.

- **Shopping centres** represent the chain stores, of medium size, that are located in modern shopping centres. Many of the stores feature specifically designed lighting in order to highlight products and displays.

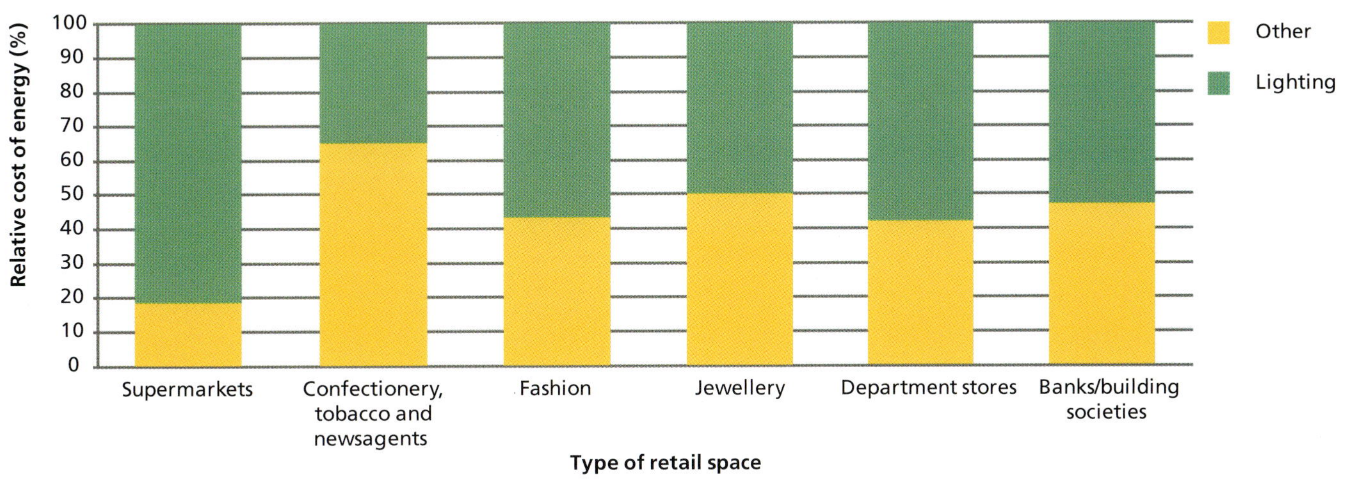

Figure 11: Lighting as a proportion of energy costs for different types of store[2] (data from EEO, 1997)

- **High street stores** represent traditional high street style stores. These are typically smaller, locally run businesses. The majority of stores were not part of a chain or brand. These stores would generally not have specialist lighting design.
- **Supermarkets** representing the major companies in the supermarket sector were also surveyed.

The survey covered 89 premises in total, with a total floor area of over 40,000 m^2, featuring over 20,000 lamps.

Using estimates of floor area for each of the above sectors, a UK-wide picture of lamp stock proportions was produced both for each sector, and in total. This gives the number of lamps of each type in use.

3.2.1 Results

UK estimate overall

The proportional results for overall stock (Figure 12) showed that fluorescents represent the majority of lamps, with approximately 64% of the stock. There are approximately twice as many T8 fluorescents as T5 fluorescents. Very few T12 fluorescents are still in use. Tungsten halogen lamps have the next largest share, with nearly 12% of stock, and then compact fluorescent (CFL) with 11%. The stock of metal halide lamps seems to be split evenly between higher wattage, higher output sources and smaller, lower wattage compact sources. These lamps represented about 5.5% of the stock each. The stock of LED lighting was approximately 1.5% in 2011, although this is expected to rise in future years.

In order to gain an understanding of the way lighting is used in different retail premises, the results are split according to the sectors described above.

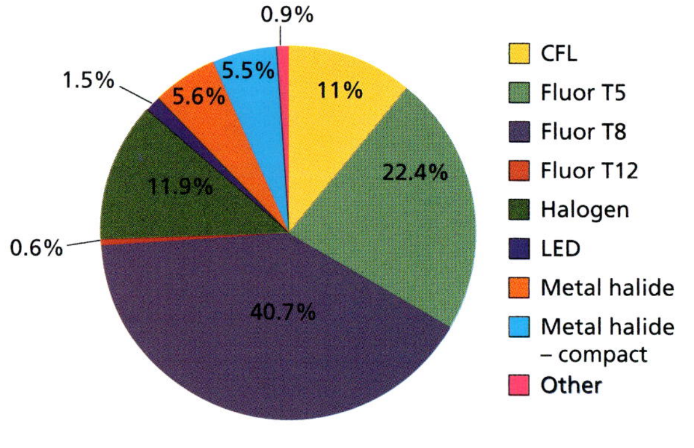

Figure 12: Estimate of proportion of UK stock in retail lighting

Shopping centres

The split of stock between the main types of lamp used in shopping centre style stores is relatively even, although obviously the wattages vary by lamp type (Figure 13). The lamp with the highest stock proportion is T8 fluorescent (19.2%), closely followed by tungsten halogen (18.4%), compact metal halide (17.5%), metal halide (16.2%) and CFL (13.6%). The more efficient T5 fluorescent is estimated to represent 12.7% of the lamp stock for these stores. There is a small proportion of LED (1.6%), but this is likely to increase rapidly over the next few years as LEDs become more widely used. The results reflect the fact that many of these stores have lighting designed to highlight products and displays. Halogen and metal halide lamps are commonly used for this task.

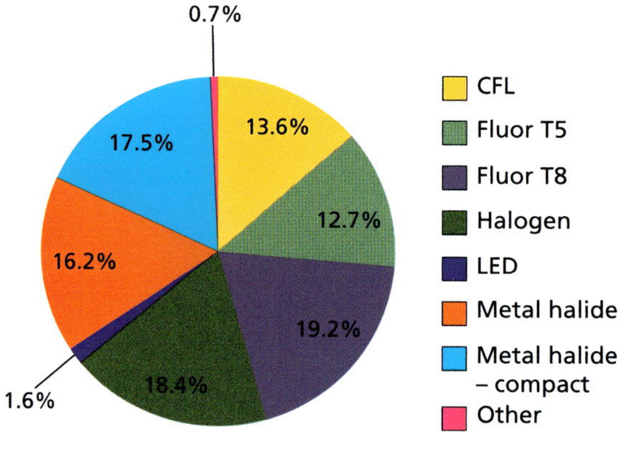

Figure 13: Estimate of proportion of lamps in UK shopping centres

High street

For the traditional high street style stores (Figure 14), the lighting stock is dominated by fluorescents. Just over half of lamp stock in use is T8 fluorescents, and just under a quarter is T5 fluorescents. The remaining quarter is made up of tungsten halogen (11%), CFL (9%), LED (1.8%) and others. The large proportion of fluorescent use is due to the focus on general lighting in these smaller stores, where there is less emphasis on display lighting than in the shopping centre style stores.

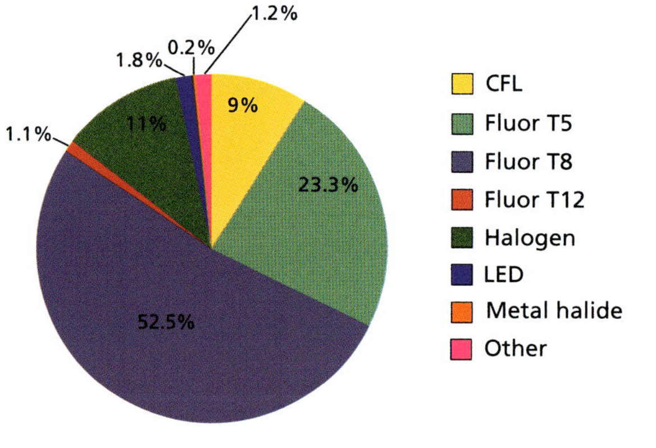

Figure 14: Estimate of proportion of lamps in UK high street shops

Retail parks

Retail park stores (Figure 15), which are closer in style to warehouses, are dominated by higher-output lamps: 42.7% of the lamps are T8 fluorescents, 38.5% are T5 fluorescents, and only 2.4% of the stock is metal halide, but these tend to be very high wattage lamps. For retail park stores the choice of general lighting seems to be between high-bay fluorescents, with each luminaire containing several lamps, and a small number of high-output metal halide lamps, with each luminaire containing one lamp. The remaining proportion of lamp stock is used for accent and display. This is mostly CFL (15.7%), but the survey suggested a small proportion of compact metal halide sources (0.6%).

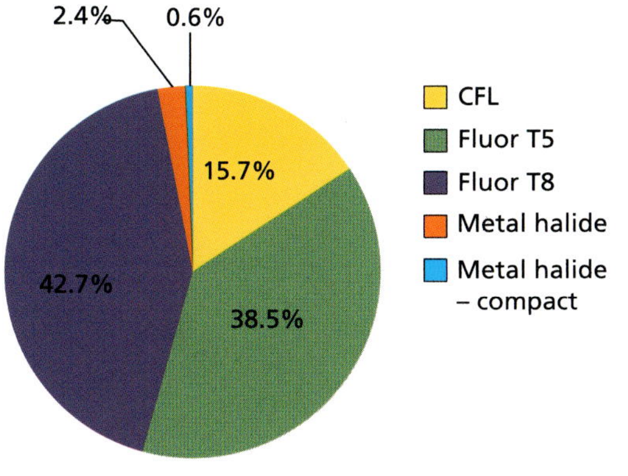

Figure 15: Estimate of proportion of lamps in UK retail parks

Supermarkets

The final category of retail store the survey investigated was supermarkets (Figure 16). The general lighting has to illuminate a large area, but there is also a focus on product displays. Fluorescent T5 are the dominant lamp type for supermarkets, with 52.1% of the lamp stock. Fluorescent T8 are the next highest in number, with 35.9%. These are not used just for general lighting, but also for display lighting, for example in refrigerated shelving. Accent lighting with CFL accounts for around 5.5% of the stock, and metal halides (5.1%) are used to highlight products, for example in the fruit and vegetable section. LEDs are being used (1.4%) for refrigerated displays, and this is predicted to increase, as LEDs replace fluorescents for this task.

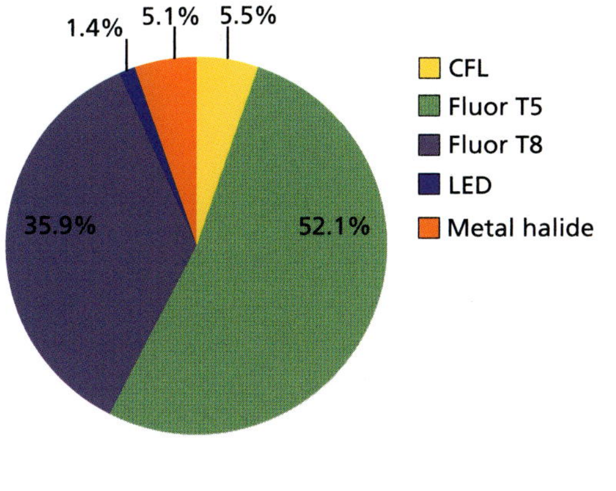

Figure 16: Estimate of proportion of lamps in UK supermarkets

3.2.2 Comparisons

The above data are based purely on the number of lamps in use. Lamps will vary in wattage, and fewer lamps may be used because they have a higher wattage (eg for retail park store metal halides), and/or are more efficient. In order to compare the different store types by including the wattage of the lamps, the power density was calculated. Power density is a measure of power use per unit area, with units of W/m^2. Each individual premises' power density is calculated from the total wattage divided by the total floor area; the results are then averaged for each retail sector. The wattage is just that of the lamps, and does not include any ballasts or controls. It is therefore not an indication of the true energy usage of the lighting installations, but is used as a comparison of lamp types and wattage between the four retail store types to get a better understanding of how lighting varies between them.

Retail park stores have the lowest average power density of 16.3 W/m^2. This is not surprising, since these relatively larger spaces can be lit more efficiently using higher-output lamps. As discussed above, high-bay metal halides and high-output fluorescents were used for general lighting.

Supermarkets have an average power density of 22.3 W/m^2. As with retail park stores, supermarkets are large areas that can be lit with higher-output lamps. The results also suggested that the more efficient T5 fluorescents were commonly in use. Supermarkets also rely on display lighting for fridges, and for areas such as the fruit and vegetable section. This requirement increases the power density relative to retail park style stores.

The high street stores surveyed have an average power density of 27.8 W/m^2. These are typically small stores. Relatively inefficient halogen lamps are still a significant proportion of lamp stock.

Shopping centre style stores have the largest average power density, at 49.5 W/m^2. This is perhaps unsurprising, given the need for bright, attractive display lighting for these stores. The survey suggested that a significant proportion of inefficient halogen lamps are being used for both display and general purposes. However, the range of lamps, and therefore of power density, did vary, depending on the products the store sells. For example, jewellery and watch stores had the highest average power density by a considerable margin, at 113 W/m^2. This suggests that shopping centre stores may have the greatest variation in power density, as it depends on the lighting design, which is often specific to the store.

John Lewis at Home, Ipswich, England

4 State-of-the-art technologies in lighting

4.1 Light sources

4.1.1 Tungsten halogen lamps

Tungsten halogen lamps are incandescent lamps in which the tungsten filament is sealed in a quartz capsule containing a small quantity of halogen. This allows the lamp to burn hotter and more efficiently, producing whiter light. Extra-low-voltage (ELV) versions are available in 6 V, 12 V and 24 V forms and require transformers, preferably electronic units that can be dimmed. The luminous efficacy is typically in the range 10–25 lm/W.

The reflector may be metal, reflecting both light and heat forwards, or dichroic, reflecting light forwards but allowing heat to pass through to the back of the luminaire. Dichroic lamps are therefore well suited to display lighting, where heat could damage the item on display, but they require suitable lampholders and luminaires. They are generally unsuitable for use in recessed lighting, as they may be an ignition source for fires.

The small size of the halogen capsule makes these lamps more suitable for directional lighting applications, such as task and accent lighting. Although tungsten halogen lamps have excellent colour rendering and a warm colour appearance, and are fully dimmable, they are inefficient. More efficient LED replacement options are already available on the market.

Tungsten halogen lamps also have a relatively short life, typically 2000 hours, which corresponds to 6–8 months in a typical retail environment. They therefore require constant replacement.

4.1.2 Fluorescent lamps

Linear fluorescent tubes and CFLs come in a range of colours, from warm appearance (2700 K) to cold daylight appearance (6500 K).

A ballast is needed to start the lamp, and to ensure a precise flow of electrical current through the tube. High-frequency electronic ballasts not only use less than half of the electricity required by conventional magnetic ballasts, but also reduce lamp energy consumption, eliminate flicker and noise, and improve lamp life.

Linear fluorescent lamps are highly efficient, with a typical luminous efficacy in the range 50–104 lm/W, and long life (typically up to 24,000 hours). Older, halophosphate lamps (being phased out under European legislation) use a single type of phosphor that degrades over the lamp's life, resulting in loss of light output, and have low colour rendering. The newer, multi-phosphor coating lamps are more efficient, with excellent colour rendering.

Older-style T12 (38 mm diameter) fluorescent tubes (also being phased out) are less efficient than the T8 (25 mm diameter) fluorescent lamps, which are available in the same lengths, and therefore can be used as direct replacements for T12 lamps in the same luminaires. Newer T5 (16 mm diameter) fluorescent lamps need different fittings.

T5 lamps are the most efficient option for enclosed luminaires. The narrow tube means that luminaire optics can be more efficient, and the T5 lamp is at its optimum performance at a higher temperature, as is commonly found inside an enclosed fitting. However, many retail luminaires, especially in supermarkets, are generally more open with less tight optical control, negating the benefit of the T5 lamp. The lamp will tend to run cooler, below its best operating temperature. So in an open luminaire a T5 installation may not offer any savings over a T8 one, and may even consume more energy. However, T5 lamps offer additional benefits. As well as the compact, neat appearance of the lamp, it is more efficient in the shorter lengths, which makes it a good choice for standard metric ceiling modules (eg 600 mm^2).

The compact fluorescent lamp (CFL) is a miniature version of the standard fluorescent tube, available in a variety of sizes and shapes. It was designed to replace the less efficient tungsten lamp, as it uses 75% less energy for the same light output (typical luminous efficacy 20–88 lm/W), and reaches longer life (typically 8000–15,000 hours). There are two types of CFL available on the market: plug-in CFLs with simple bayonet or Edison screw caps, which have an integral ballast and can be used as direct replacements for tungsten lamps in most existing luminaires, and pin-base CFLs, which need separate control gear and specially designed luminaires.

Most 4-pin fluorescent tubes and 4-pin-base CFLs can be dimmed, but it is not always possible to dim smoothly down to near extinction. With electronic dimming ballasts it is possible to dim some fluorescent lamps down to 2–5% of their nominal light output. A small number of plug-in CFLs incorporate special electronic gear that allows dimming, but most plug-in CFLs cannot be dimmed.

4.1.3 Metal halide lamps

Metal halide (MH) lamps are high-pressure mercury discharge lamps that produce high-quality white light with good colour rendering and low energy consumption. Newer technology ceramic arc-tube versions offer excellent colour rendering, and have less light loss over time and a longer lamp life (up to 20,000 hours).

In larger, shed-type retail spaces the choice is often between fluorescent lighting and metal halide lamps in high-bay fittings. High-bay metal halide is usually cheaper to install, and gives an 'industrial' feel to the space, which suits some warehouse-type applications. Fluorescent tubes give a more uniform lighting, with less glare. They are often installed in multi-lamp luminaires, which means that if a lamp fails, the others will keep working and the failure will be less noticeable. Failed lamps in metal halide luminaires are easy to spot, and may give rise to gloomy areas in the space, especially if there is high racking. Finally, fluorescent lighting is the preferred option in rooflit spaces, as it can be readily dimmed or switched off on bright days.

Compact metal halide (CDM) lamps, based on ceramic technology reach high luminous efficacy values, typically in the range 65–110 lm/W, which makes them an excellent alternative for spotlighting. However, they require separate control gear, and it takes several minutes from start-up to full output. They cannot be restarted from hot when switched off: therefore these lamps are most suitable for continuous lighting. Compact metal halide lamps can typically be dimmed down to 50% of the nominal light output by the use of special dimmable control gear and compatible dimmers. Some CDM lamps have had maintenance and installation problems, because the end of the lamps (GU8.5 format) is not robust enough; if it is moved from side to side in its fitting the wires at the base of the lamp break, and it has to be thrown away. New versions are now being produced with a more robust end cap.

4.1.4 Induction lamps

Induction lamps are gas discharge lamps based on fluorescent technology that do not have electrodes, but use an induction coil to induce the electric field in the discharge tube. The induction coil in the centre of the lamp is fed from a high-frequency generator, which is the external control gear required by these lamps.

Induction lamps have many of the same properties as fluorescent lamps, but are slightly less efficient, with lamp efficacy typically in the range 47–80 lm/W, and cannot be dimmed. However, because they do not have electrodes to fail, induction lamps have a significantly longer life (up to 100,000 hours), which makes them an ideal option for locations where lamp replacement is difficult, such as high-bay retail spaces.

4.1.5 LEDs

LEDs can help achieve energy-efficient lighting in retail spaces, especially by replacing inefficient tungsten halogen lamps used for accent and display lighting (Table 2). Combining LEDs with appropriate controls that include daylight-based dimming where applicable can provide substantial energy savings up to 80–90%[14].

LEDs are very small semiconductor diodes that emit coloured light when a low-voltage DC current is applied. White light can be produced either by combining red, green and blue LEDs, or by using a phosphor coating to convert UV or blue light to white. LEDs are inherently directional light sources, suitable for decorative, accent and display lighting. They have very long life (up to 50,000 hours) and fast-improving luminous efficacy (currently on average 40–90 lm/W).

However, LEDs perform better at low temperatures; they tend to become less efficient when hot, and their colour quality and lifetime may also deteriorate. They therefore need good heat-sinking or ventilation. Specialist luminaires with appropriate optics and heat sinks are preferred. Retrofitting an LED lamp into an existing fitting may not work as well as expected, particularly if the existing fitting is small and enclosed, or has insulation around it (for example a downlight in an insulated ceiling). However, the good performance of LEDs at low temperatures makes them ideal for use in refrigeration units.

LEDs use an electronic driver to convert the power supply into the required voltage and current for the device. This may be integrated into the LED lamp, or a separate product. LEDs are normally straightforward to dim, and unlike other lamps their efficacy can improve when they are dimmed, as the lamp is running cooler. Dimming equipment must be compatible with LED lighting.

CASE STUDY Benetton , Padua, Italy

The LED-only scheme[16] (Figure 17) delivers an average light level of around 500 lx on the walkways and a light level on specific focal areas of the merchandise of around 900–1000 lx. The LEDs are all 16 W (50 lm/W), 3000 K and CRI 90. The installed lighting load is 17 W/m^2.

Figure 17: Fully LED-lit clothes store (© Reggiani)

A wide variety of coloured LEDs are available for decorative effects (Figure 18). Using coloured LEDs is generally much more energy-efficient than using another lamp type with a coloured filter. Coloured lighting should be used sparingly, as it can make the goods on sale, and the shoppers themselves, look strange.

Some LED types can change colour as well. This can be used in a dynamic way to add interest to the decorative effects created. Lighting can be varied fairly rapidly, so that the colour change is obvious; or it can be varied very gradually, for example by changing from cooler colours during the day to warmer, 'cosier' lighting at night. Alternatively it can be used to generate a constant light that is specifically tailored to the products on display, for example light with a higher red content for meat counters. The colour balance of these LEDs can change over their lifetime; high-quality products will incorporate special circuitry to balance the colour output over time.

LEDs vary widely in their performance, and the manufacturer should provide detailed information about the lamp output; sometimes LEDs that are sold as equivalent to tungsten halogen lamps actually produce less light.

4.2 Ballasts and control gear

The power a lamp consumes (in Watts) depends on the type of lamp, and also on the ballast or control gear. Fluorescent and discharge lamps require a starter and ballast to control the current through the lamp. The least efficient ballasts are now banned by European legislation[17]. The older, magnetic wire-wound ballasts may typically consume around 20–25% of rated lamp power. Electronic ballasts operate the lamp at high frequency. This improves the efficiency of the lamp, so that the same light output can be achieved with a lower lamp

Figure 18: Colour-changing LEDs create attractive decorative effects in the retail environment (© iGuzzini)

Table 2: Summary of key characteristics for different lamp types (sample images © Osram)

Lamp type	Sample image	Luminous efficacy (lm/W)	Colour appearance (K)	Colour rendering (CRI)	Lamp life (hours)
Tungsten halogen		10–25	2700–3000	100	1500–2000
T8 linear fluorescent		50–96	2700–6500	50–98	8000–24,000
T5 linear fluorescent		80–104	2700–6500	82–95	8000–24,000
Plug-in compact fluorescent		20–74	2700–6500	80–90	6000–12,000
Pin-base compact fluorescent		30–88	2700–6500	85–90	8000–15,000

Table 2 (contd): Summary of key characteristics for different lamp types (sample images © Osram)

Lamp type	Sample image	Luminous efficacy (lm/W)	Colour appearance (K)	Colour rendering (CRI)	Lamp life (hours)
Metal halide (quartz) (MH)	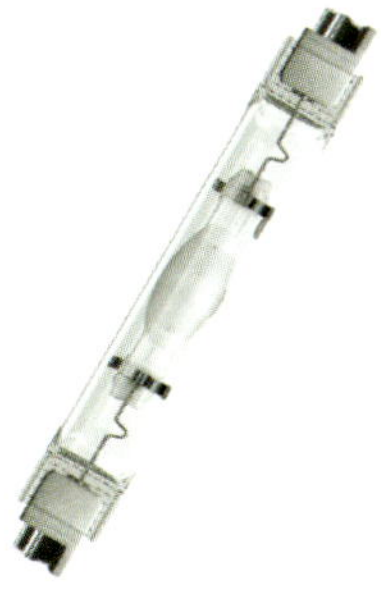	60–118	3000–6500	60–90	2000–20,000
Compact metal halide (ceramic) (CDM)	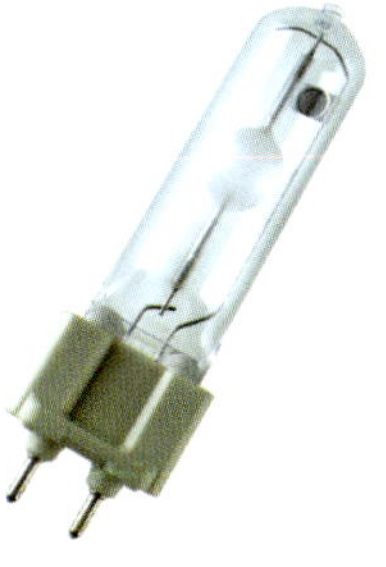	65–110	3000–4400	70–93	6000–24,000
Induction		47–80	2500–4000	80	60,000–100,000
LED		40–90	2700–6500	60–90	10,000–50,000

power consumption. Typically, the total circuit power of a high-frequency circuit is around 25% less than that of a lamp with a magnetic ballast giving a similar light output.

High-frequency electronic ballasts are also lighter in weight than their electromagnetic equivalents, are silent in operation, and switch the lamp off automatically at the end of its life, eliminating flashing. They may, however, have shorter lives themselves, owing to the inclusion of poor electronic components in some units. The high-frequency operation eliminates flicker, and has been shown[18] to reduce the reported incidence of headaches and eyestrain in some individuals.

In new stores owned by the major chains, high-frequency ballasts are generally used for fluorescent lighting. Electronic control gear is also available for the low-wattage CDM metal halide lamps. Some retailers favour it, because it gives improved lamp life (important because of the high costs of CDM lamps) with lower lumen depreciation.

Tungsten lamps do not require a ballast. However, low-voltage tungsten halogen lamps will need a transformer.

4.3 Luminaires/light fittings

There is a large variety of fittings to provide different types of lighting, and to suit different lamp types. The most common luminaire types used for lighting in retail spaces are listed in Table 3.

The light distribution of direct luminaires is predominantly downward, and therefore these fittings incorporate various forms of optical control and antiglare shielding, such as diffusers, parabolic reflectors or louvres. Direct luminaires are the most energy-efficient fitting type, but produce dark ceilings and high contrasts in the lit environment, with poor illuminance uniformity in obstructed spaces. This effect can be overcome to an extent by using high-reflectance surfaces for interiors.

Indirect luminaires provide high illuminance uniformity and bright appearance of the space, excluding the risk of glare, but have the lowest energy efficiency for lighting. They are most practical where the ceiling height is at least 2.8 m.

Direct/indirect luminaires combine upward and downward light distribution, and use different optical controls for these components. Their energy efficiency is higher than that of indirect luminaires, and they provide a brighter ceiling and better uniformity than direct lighting.

Downlights are direct luminaires with small light-emitting apertures of various beam spreads and sizes, which are meant to make the lighting system inconspicuous. Because of the predominantly downward light distribution, they create dark ceilings and high contrasts, and, being ceiling-recessed, they require extra care to prevent them from becoming a fire hazard for insulated ceilings.

The choice of luminaire can have a significant effect on the energy consumed. The efficiency of a luminaire is quantified by the light output ratio (LOR) (Figure 19). This is the ratio of the light emitted by the luminaire to the light output of the lamps in it. In practice, the LORs of different ranges of luminaires can vary considerably, even for luminaires that look similar. The LOR depends on the quality of the materials used, as well as on the basic design of the luminaire. Most luminaires for general lighting would be expected to have an LOR of 0.6 or more, but they typically vary from 0.4–0.8, and can be better or worse than this range. The cheaper luminaires tend to have lower LORs. Although it is important to have a high LOR for good energy efficiency, it is equally important to ensure that the light is adequately directed to where it is needed. For example, bare batten fluorescent tubes have high overall LOR values, but the light is equally distributed in all directions, giving a flat illuminance to all surfaces, rather than concentrating the light on the goods or work area of interest.

In fact, few fluorescent luminaires appear to be specifically targeted at the retail market, with its need for a spread of light to illuminate vertical surfaces; recessed downlighting and industrial-type troughs are the commonest types of fitting. The light reflectance of the vertical surfaces themselves may also have an impact on the effectiveness of the lighting. In a retail shed, where stock is piled high on vertical racks, the lightness of the packaging has an impact on the illuminance at the bottom of the aisles.

Spotlights are directional fittings with narrow beams, typically in the range 5–30°. They can use mains voltage or low-voltage lamps; in the latter case, transformers (typically accompanying the fittings) are required. Spotlights have reflectors for optical control, and because of to their directional character they can cause glare.

For spotlights and projector lamps, the beam spread is also important. Light is projected through a defined cone, which depends on the reflector in the lamp or fitting. This can be quantified by specifying the beam angle of the lamp and its intensity in a particular direction (measured in candelas). Alternatively the amount of light (in lux) falling on a surface at a given distance can be measured (1 lx = 1 lm/m^2). For a point source, the illuminance at 1 m distance is equal to the intensity in candelas, and the illuminance drops with the square of the distance from the lamp. Thus the illuminance 2 m from the lamp will be a quarter of its value at 1 m from the lamp.

In a lamp catalogue, the intensity quoted is generally the intensity of the peak area at or around the centre of the beam. Similarly, the illuminances quoted are usually those at the centre of the beam.

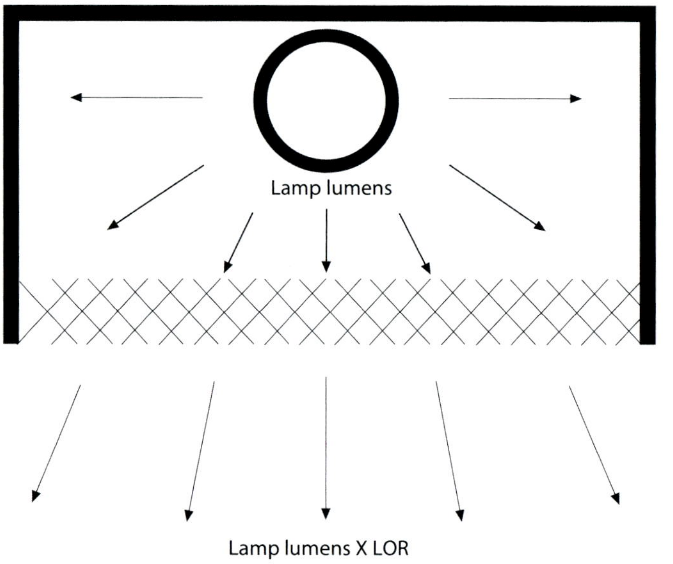

Figure 19: The light output ratio of a luminaire is the light coming out of the fitting divided by the light from the bare lamp(s)

Table 3: Summary of main luminaire types using efficient light sources in retail lighting

Fitting type	Sample image	Typical fixing	Purpose	Efficient lamp types
Downlight	 © Philips Lighting	Ceiling-recessed	General lighting or accent lighting (adjustable aiming)	CDM, CFL, LED
Direct luminaire	 © Philips Lighting	Ceiling-recessed, ceiling-mounted or suspended below ceiling	General lighting	Linear fluorescent, CFL, MH, LED
Indirect luminaire	 © Erco	Suspended below ceiling, or wall-mounted	General lighting	Linear fluorescent, MH, LED
Direct/indirect luminaire	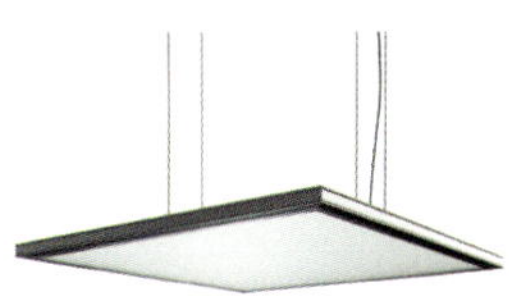 © iGuzzini	Suspended below ceiling, or wall-mounted	General lighting	Linear fluorescent, CFL, LED
Spotlight	 © Philips Lighting	Mounted on lighting track or baseplate	Accent or display lighting	CDM, LED
Internal floodlight	 © Erco	Mounted on lighting track, ceiling- or wall-mounted	Display lighting	MH, LED

Table 3 (contd): Summary of main luminaire types using efficient light sources in retail lighting

Fitting type	Sample image	Typical fixing	Purpose	Efficient lamp types
Wall-washer	 © Robus	Ceiling/ground recessed or mounted on lighting track	Ambient, accent or display lighting	MH, linear fluorescent, LED
Decorative	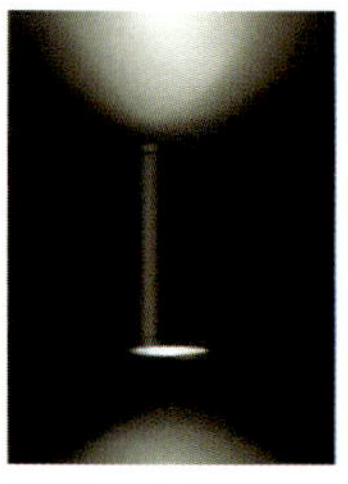 © iGuzzini	Wall-mounted	Architectural lighting	LED, CDM, CFL
Fibre optics	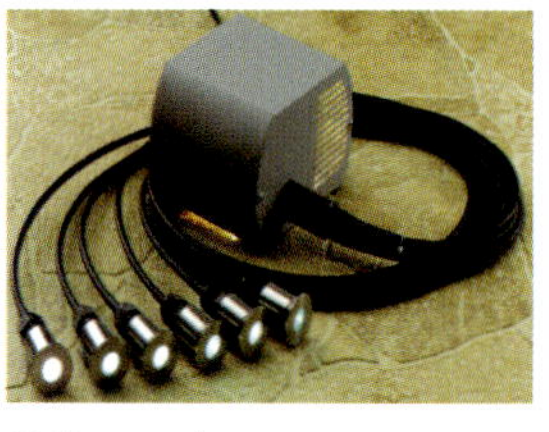 © Crescent	Ceiling-recessed, wall-mounted	Architectural lighting	MH, LED
External floodlight	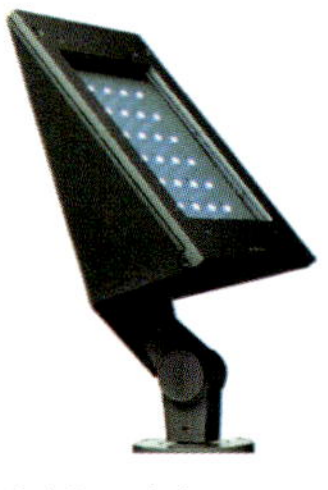 © iGuzzini	Wall-mounted or ground-recessed	Architectural lighting	MH, LED
Lantern, bollard, wall-mounted downlighter and ground-recessed spotlight	 © iGuzzini	External walls, other external areas	Outdoor lighting	MH, CFL, LED

CASE STUDY Tesco, Barnstaple, North Devon

The installation of pre-wired, daylight-based, dimmable 2 × 73 W T5 Eco fluorescent fittings, with adjustable reflector and louvre for precise angling to the point of sale (Figure 20), provided a 28% energy saving, 30% reduction in the number of lamps required, and 25% reduction in installation time compared with the previous lighting scheme[19].

Figure 20: Daylight-based dimmable T5 fluorescent fittings with adjustable reflector and louvre produce the right light output at the level of the merchandise (© Whitecroft Lighting)

The beam angle is the width in degrees of that part of the beam for which the luminous intensity is at least half the maximum intensity. So at half this angle from the centre of the beam, the luminous intensity and illuminance will be half their maximum values. For typical interior display lighting, lamps with beam angles below 20° are usually classified as spotlights and those over 20° as floodlights, but the exact nomenclature varies with product and manufacturer. Downlights generally use a wide beam angle, of 50–80°.

Even if different lamps have the same intensity and beam angle, it can still be hard to compare them. Some lamps have a very peaky distribution, with the intensity dropping substantially away from the maximum. Others give a much more uniform pool of light. Manufacturers may supply a polar curve: a graph that gives the intensity of the lamp at different angles.

A narrow beam angle allows the spotlight to be more noticeable, as the illuminance it creates within the cone is larger (Figure 21). This strategy can allow a higher-wattage, wide-beam fitting to be replaced with a narrower-cone reflector with the same degree of 'punch'. Of course, the area lit is smaller, so this approach is best suited to spotlights with a relatively high mounting height, as often occurs in retail sheds and supermarkets. For example, a large supermarket used 70 W CDM-T lamps with a 40° beam angle, mounted about 4 m above the floor. Changing these to a 35 W version with a 20° beam angle halved the power used, and gave increased emphasis on the displays.

Circular or rectangular floodlights with wider beam angles and incorporating MH or LED lamps can be used to light larger indoor display areas, typically to a uniform high light level. Flexible solutions using adjustable or track-sliding floodlights help meet specific requirements for various merchandise display arrangements.

Wall-washers are luminaires with asymmetrical light beams that are used to 'wash' large areas evenly with light. Careful aiming is required to achieve best effects.

Sometimes perimeter ambient lighting is achieved by mounting fluorescent or LED lamps in open coving. This can be inefficient; in one store, lamps were installed in an open cove that allowed light spill into the false ceiling cavity (the cove needed to be kept open for ventilation purposes). Such light spill could be avoided by the use of fittings with appropriate optical design to orient light where it is needed (ie asymmetric reflectors). This can also give extra 'punch' and emphasis to the lit effect (Figure 22).

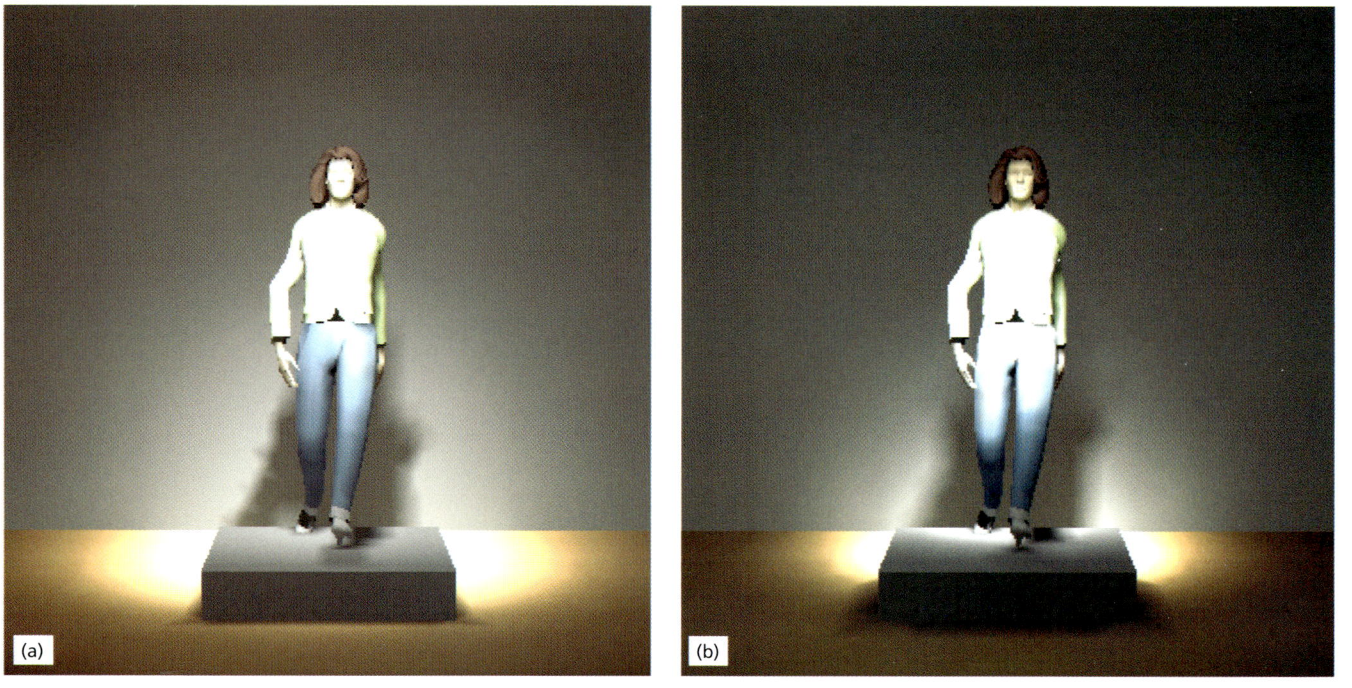

Figure 21: (a) Light effects for a 70 W CDM spotlight of 28° beam angle compared with **(b)** a 35 W CDM spotlight of 10° beam angle. Energy savings of 50% can be achieved by aiming the light on the mannequin, and reducing background lighting. This also increases the emphasis on the display.

Figure 22: Perimeter lighting by using LEDs in a cove arrangement that can project the light onto the walls or the ceiling (© Philips Lighting)

Interesting architectural features such as wall structures or columns can be enhanced by appropriate lighting, improving the space appearance and setting a more attractive environment for shoppers. In most cases a store's character, image and brand can be effectively defined by architectural lighting, with emphasis being placed on entrances. Given the smaller size and continuously increasing luminous efficacy of LED lighting, it can provide very discrete and flexible solutions that are ideal for architectural lighting.

Fibre optic systems can be used to distribute light from one large, energy-efficient light source to multiple, smaller terminations from which light can be directed as required. The advantages of using fibre optic systems are flexibility, the use of efficient light sources, cool operation, no need for electrical wiring in ceiling voids or wall cavities, and easy maintenance. However, a bespoke lighting design is typically required for each application, and all lights switch off upon lamp failure[20].

Fibre optic systems operate based on total internal reflection of light within the fibres. In this way, light emitted by a high-output lamp (called a light generator) is transported through the fibre optics to the terminations. The absence of heat and the elimination of electric wiring are further benefits of using fibre optics. Because they do not cause heat build-up, fibre optic systems are commonly chosen for showcase lighting.

External floodlights can be used to wash large areas with light, or to highlight specific architectural features of buildings. Typically, they are equipped with high-wattage lamps, and their light distribution can be either symmetrical or asymmetrical. Floodlights need appropriate protection against dust and moisture, and correct aiming is required to avoid glare to passers-by. In the past, coloured filters had to be used if changes in light colour were required. This can now be carried out efficiently by colour-changing LED floodlights.

Other external lighting fittings should contribute to emphasising key areas such as store entrances and exits, pathways, parking facilities and crosswalks. Specific fittings, such as lanterns, bollards, wall-mounted downlighters and ground-recessed spotlights, should incorporate energy-efficient light sources such as metal halide, compact fluorescent or LED lamps.

4.4 Daylighting

Using daylight as a primary light source can potentially save energy by reducing the need for electric lighting. However, over-glazing can lead to high solar heat gains. The aim should be to provide generous levels of daylight with reasonably low associated solar heat gains to avoid using extra energy for cooling. Daylight into a retail space needs to be controlled by a combination of glazed or reflective elements, such as high-performance windows or rooflights, or optical systems able to guide daylight to where it is needed. Lighting controls are required to adjust the electric lighting to match daylight variability.

It is also important to distinguish between direct sunlight and diffuse skylight. Sunlight gives patches of high illuminance and strong contrasts, providing variation in light intensity, pattern and colour. Hence sunlight can make a retail space more vivid and attractive to customers, but appropriate shading is needed to avoid thermal and visual discomfort. Diffuse skylight is less dynamic and intense, and reduces contrasts within internal spaces, but it provides a higher uniformity of indoor daylight levels, and helps make a retail space softer and more pleasant for shoppers.

Daylight has a number of valuable visual effects. It gives a building a unique variety and interest, providing the retail space with a bright and dynamic appearance that has benefits for the mood of both staff and customers. The variation in daylight stimulates both the human visual system and the human circadian system, which regulates body rhythms and functions fundamental to health and well-being. Staff often benefit more than customers, as they are continually in the store.

The full spectrum of daylight ensures ideal colour rendering (CRI of 100), which eliminates distortion of colour perception of merchandise in any daylit sales area. This makes daylight an excellent, high-quality light source. Adequate daylighting can enhance a store's appearance, thus attracting and retaining customers and increasing sales. Different studies have been aimed at assessing daylight impacts on sales in retail spaces, and the general finding is that daylight can typically have a beneficial effect. US research studied 108 chain stores selling similar merchandise and having similar space layout. Analysis of the sales data predicted that adding skylights to the average non-daylit store in this chain would increase sales by 40%[12]. Another technical report was prepared by the same group in 2003, following a more sophisticated evaluation that was carried out over almost three years on a different chain of stores operating in a different retail sector[21]. A larger number of variables were taken into account, from monthly sales and transactions per store to corporate and census data, local market and store comfort conditions, and their interactions with the amount of daylight available in the store. The results of this study were, interestingly, quite different from the previous study, showing significantly lower mean percentage changes in sales with daylight. For that particular chain of stores, the new study showed a predicted average increase in sales associated solely with the access of daylight of 1–6%, depending on the modelling data.

> **Using daylight as a primary light source can potentially save energy by reducing the need for electric lighting**

When considering the interaction between daylight and available parking, the results were much better and showed a net increase in sales volumes. With normal parking areas, daylight can increase sales by 6%, but when parking area is greater than normal, daylight can increase sales by 20%. Furthermore, reducing parking to less than 90% of the norm seemed to lead to negative effects from daylight. The analysis suggests that adding daylight to a store would have benefits only if sufficient parking were also available.

The latter study also assessed the energy savings for lighting from daylighting and daylight-based lighting controls, found to be in the range of 20–30%. The overall energy savings for lighting and HVAC together were in the range 15–25%. These studies indicate that providing daylight in stores may typically have some positive effects on sales, but other factors that relate to customer comfort and facilities should also be taken into account.

The use of daylighting strategies can also help a store's design to earn credits under various green schemes or standards, by providing benefits in terms of reduced energy and environmental impacts, indoor environment quality, and innovation and design.

4.4.1 Daylighting strategies for retail

Daylight in stores is usually limited, owing to a range of factors. Most commonly, the building type and layout do not favour daylight penetration into internal areas. Moreover, the majority of retailers prefer to black out windows into sales areas to increase display and/or sales area, or to allow for other ancillary spaces, ie corridors or storage. The potential for sidelighting is therefore limited, owing to enclosed storefronts, or to the use of opaque display windows or light boxes for display purposes. There may be limited scope for toplighting because of the absence of roof apertures for daylight, either from early design stages or from later blackouts during store operation. External obstructions, particularly in urban environments, also limit daylight availability in retail spaces.

Where generous daylight penetration is observed, it is generally provided by completely glazed storefronts or sidewalls. The highest daylight levels were identified in car showrooms, but even in such cases there was potential for improvement, as electric lighting was linked to daylight in only a very few cases. Most of the electric lighting was kept on in daylit spaces, cancelling the benefits of daylight in terms of energy saving.

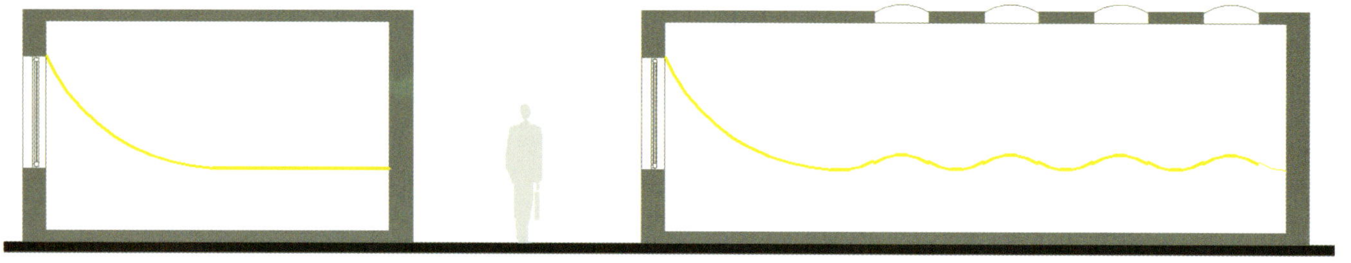

Figure 23: Daylight distribution is more uniform for toplighting than for sidelighting strategies

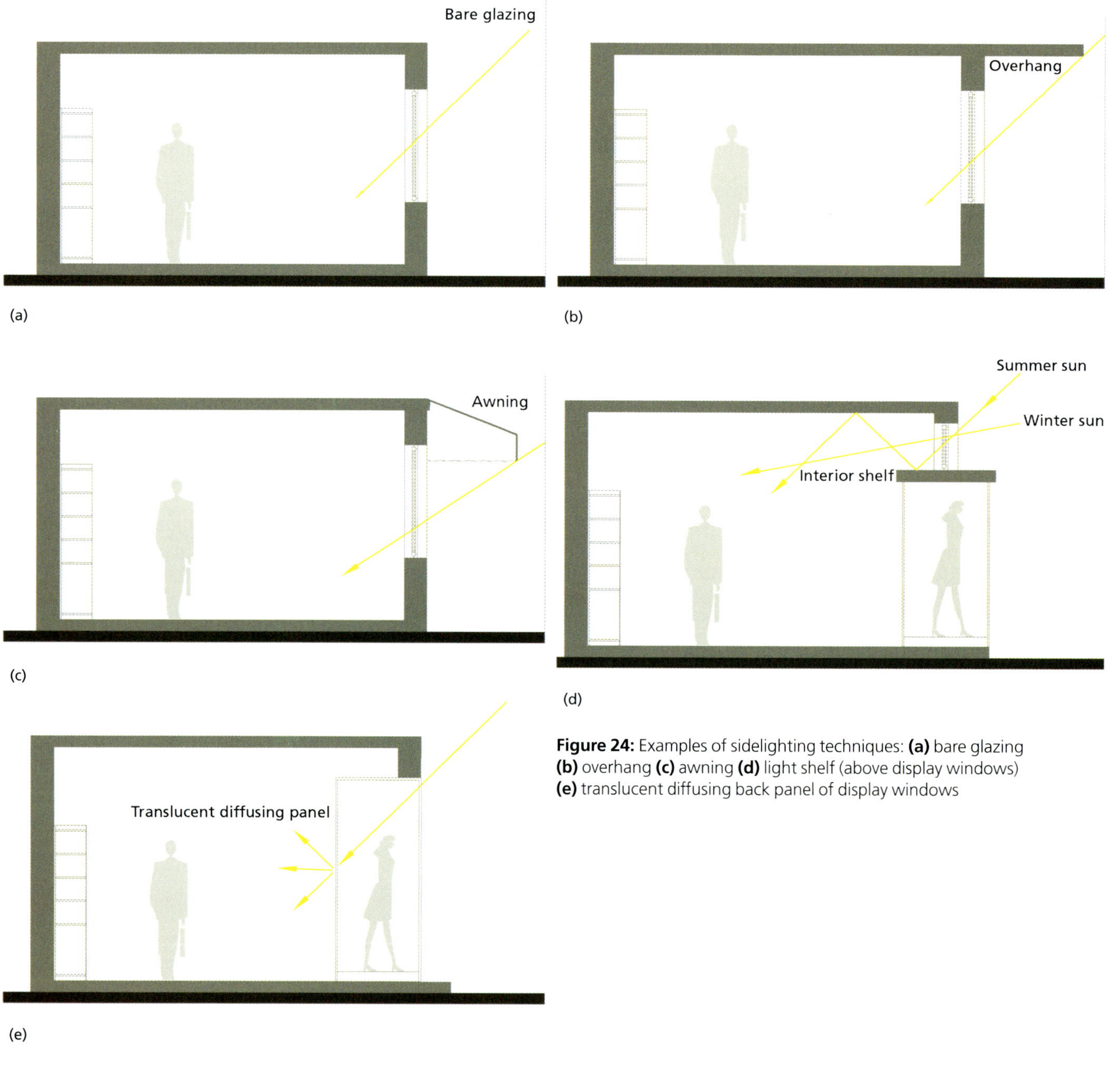

Figure 24: Examples of sidelighting techniques: **(a)** bare glazing **(b)** overhang **(c)** awning **(d)** light shelf (above display windows) **(e)** translucent diffusing back panel of display windows

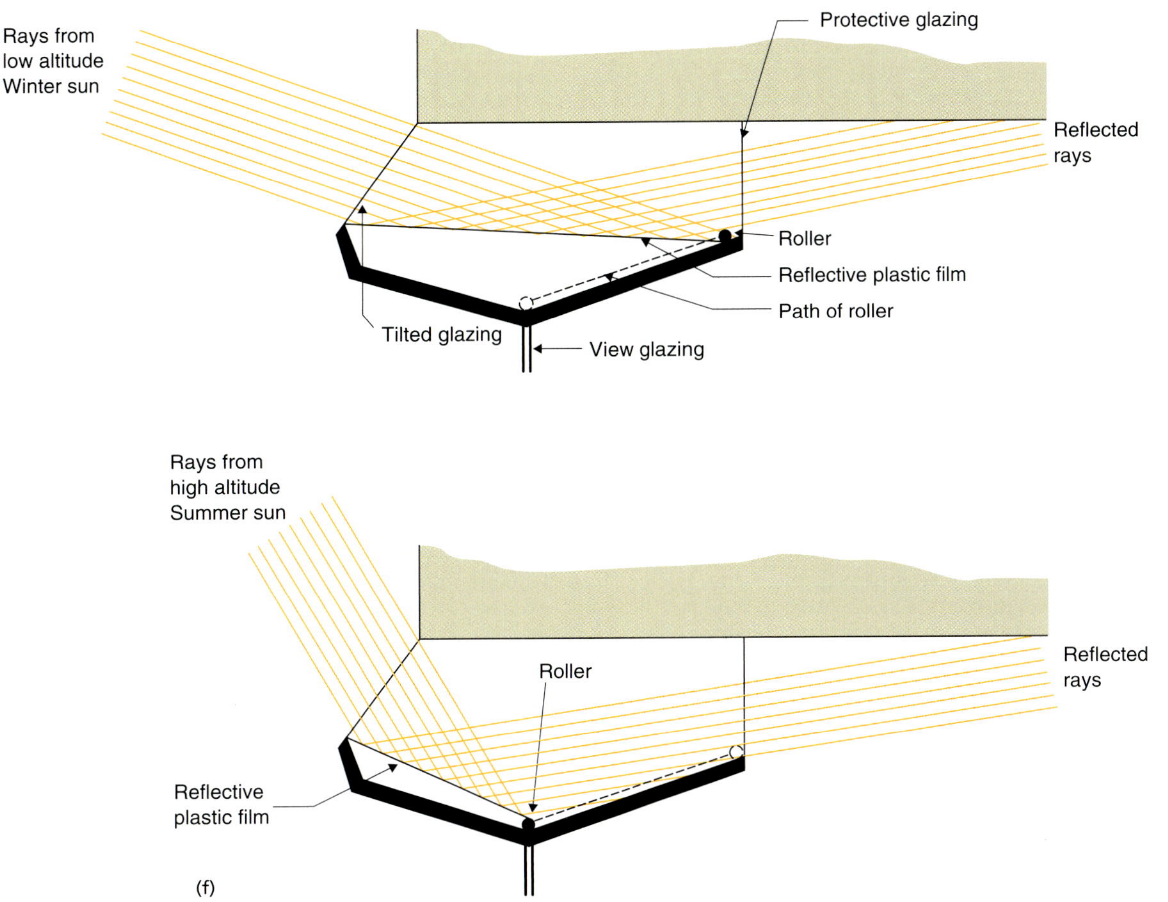

(f)

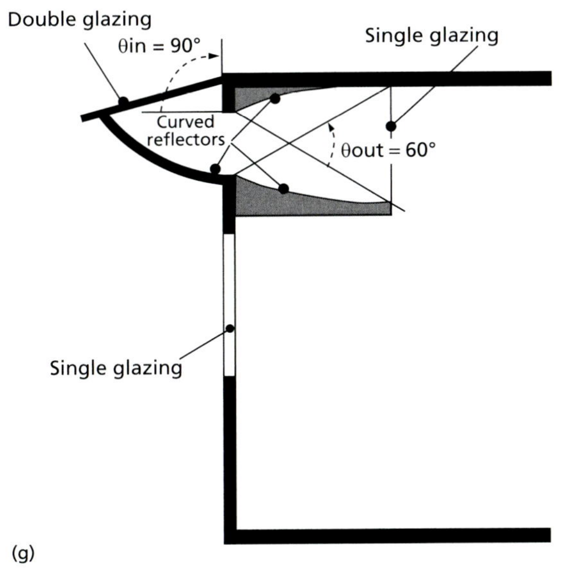

(g)

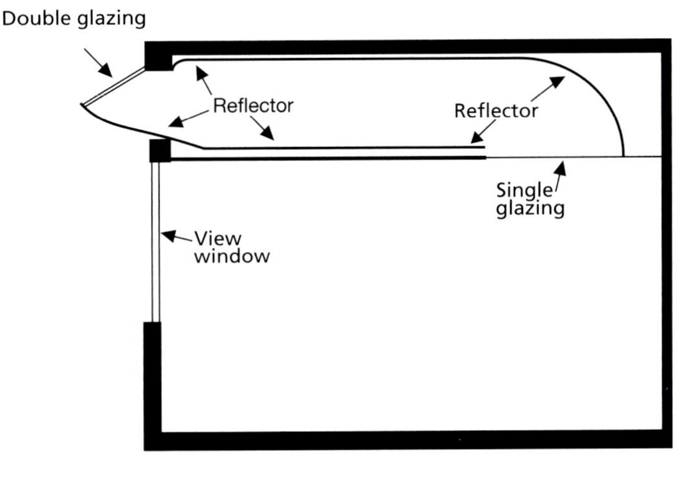

(h)

Figure 24 (contd): Examples of sidelighting techniques: **(f)** variable-area, light-reflecting assembly **(g)** anidolic reflector system **(h)** anidolic ceiling arrangement

Advanced daylighting strategies should be designed to provide sufficient daylight levels without any unwanted side effects. Heat transfer, glare control, and variations in daylight availability must be carefully balanced. Sidelighting is the most common strategy adopted for daylighting, typically in store fronts. It can provide a view out that may be appreciated by store staff, but not necessarily by shoppers, who typically spend less time in the space and are more interested in observing the merchandise. Sidelighting can be integrated in multi-storey buildings but the amount of daylight that penetrates into the space depends on the building orientation and the absence of obstructions, and there is a higher risk of glare from direct and reflected sunlight. The distribution of daylight in the space from sidelighting is uneven (Figure 23), with higher daylight levels in areas nearer to windows and decreasing amounts further away into the space. For this reason, sidelighting is not effective for retail spaces with a deep layout. Advanced sidelighting strategies (Figure 24) can improve daylight penetration into the depth of the building by redirecting it onto a reflective ceiling. Light shelves, sun-directing glass or anidolic collectors can be installed in the upper part of windows to redirect daylight deeper into the buildings[22]. The electric lighting system needs to be controlled separately in the daylit and non-daylit areas to maximise energy savings while providing the required light levels.

Toplighting strategies (Figure 25) use openings in the roof to allow daylight penetration into a space: hence their application is limited to single-storey buildings, or to the top floor of multi-storey buildings. However, daylight uniformity is significantly improved throughout the whole space, and there is less impact from obstructions, with maximum available daylight at all times. There is a limited view out, but toplighting strategies in atria and courtyards allow occupants to experience connection to the outside at all times of the day. Because of higher and longer exposure to sunlight, irrespective of orientation, toplighting strategies – other than northlights – typically incur a higher risk of overheating than sidelighting. Soft-coat, low-emissivity glazing with low solar transmittance (g-value) but higher daylight transmittance should be considered for such strategies.

Typically roof apertures over 8–10% of the whole roof area can be sufficient to keep electric lighting turned off during a significant part of the daytime. Substantial savings can be made by capitalising on the availability of natural light, while creating a pleasant, airy atmosphere that is favoured by shoppers.

Atria and light wells are still valued as architectural features. As well as bringing daylight into the building, the atrium provides a focal point, and aids customer orientation within the store. Care needs to be taken to ensure that adjoining floor areas in the store do not look gloomy, compared with the bright atrium. The following measures can help:

- Turn off the electric lighting in the atrium during the day.
- Consider stepping or staggering the atrium so that more of the sales area is directly daylit.
- Consider having an opaque roof in the centre of the atrium (Figure 27), so that light enters from the sides and goes into the sales areas, rather than the predominant flow of light being directly downwards.
- Have reflective surfaces in the atrium and adjoining sales areas to bounce light around the building.

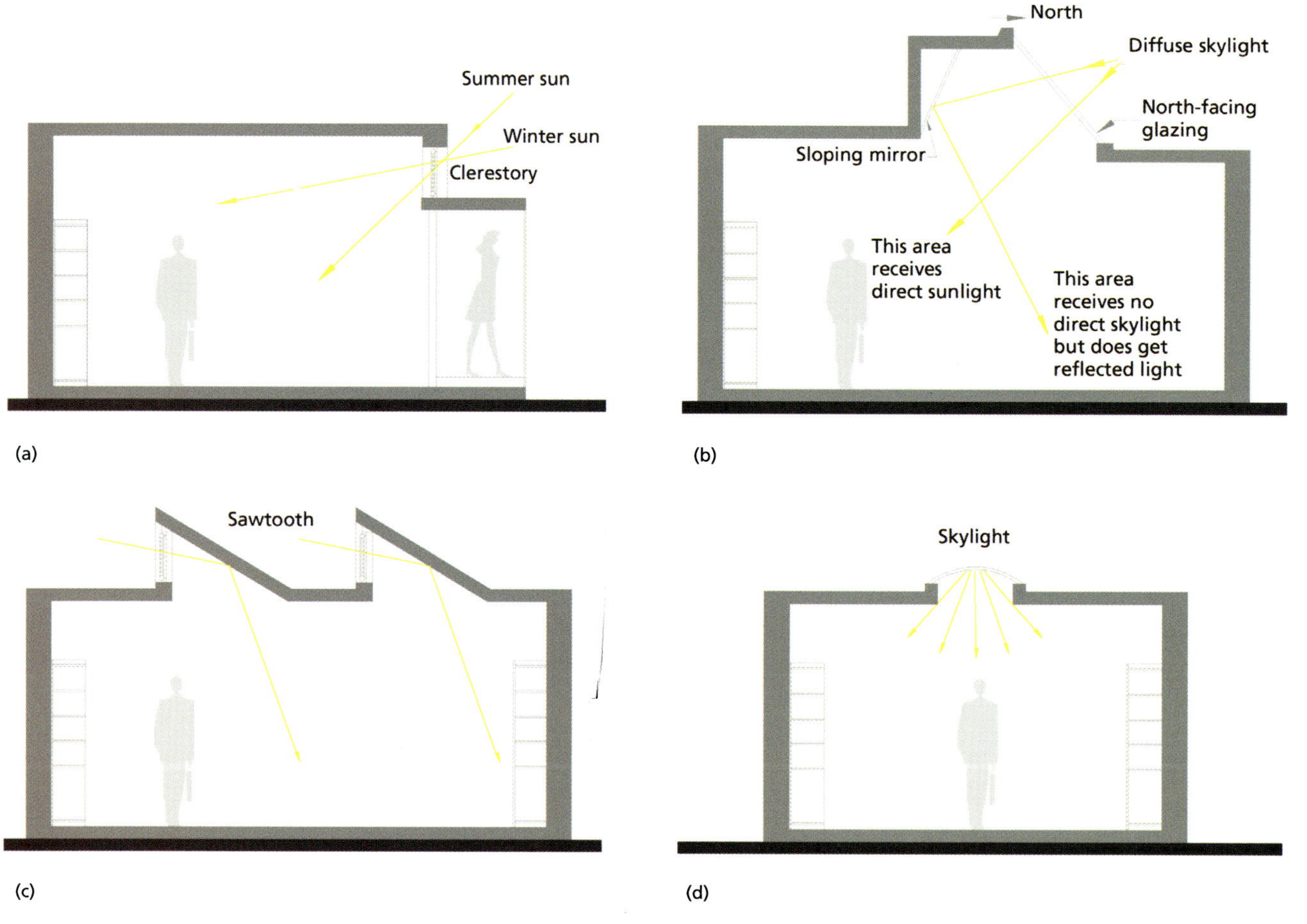

Figure 25: Examples of toplighting techniques: **(a)** clerestory **(b)** northlight **(c)** sawtooth **(d)** rooflight

CASE STUDY Selfridges, Oxford Street, London

The lighting of the Shoe Galleries in the London Selfridges Store (Oxford Street) was designed to enhance the product appearance and interior architecture (Figure 26). NDYLIGHT modelled the natural lighting potential to identify where strategic rooflights should be inserted to enable visitors to see and try shoes under tempered, natural light. The resulting effect is a dramatic presentation of products, coupled with a feeling of spaciousness and naturalness.

A combination of energy-efficient light sources, including metal halide, dimmable LED and fluorescent lamps, was used to help reduce the original installed lighting loads to approximately 23 W/m^2, which is currently well below the typical loads used in high-end luxury retail stores.

Figure 26: The shoes gallery at Selfridges: enhancement of space appearance and internal architecture by a combination of rooflights with dimmable electric lighting (© NDYLIGHT Lighting Designers and Consultants)

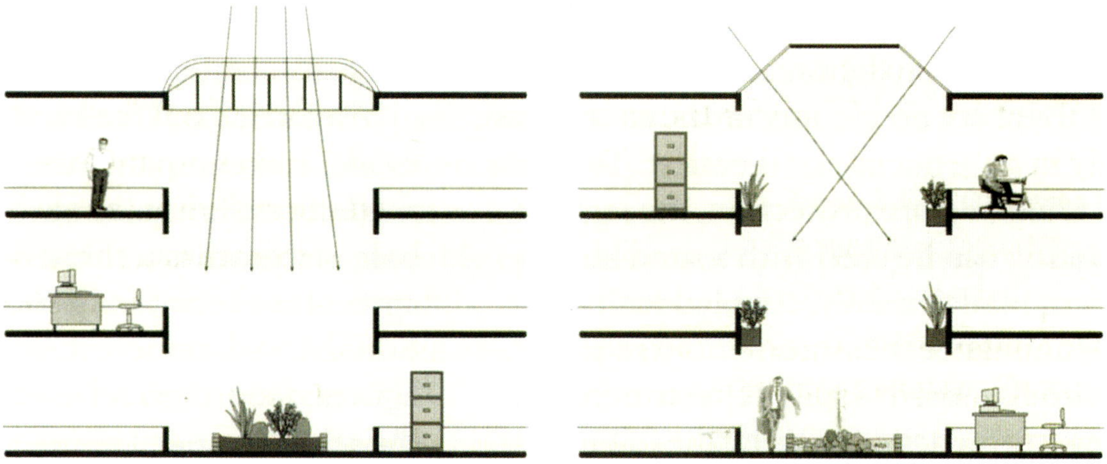

(a) (b)

Figure 27: Changing the roof profile affects daylight distribution. With the egg-crate roof **(a)** most of the light goes straight down, and less reaches the atrium sides. With sloping apertures **(b)** more light is directed to the sides, and less to the atrium floor.

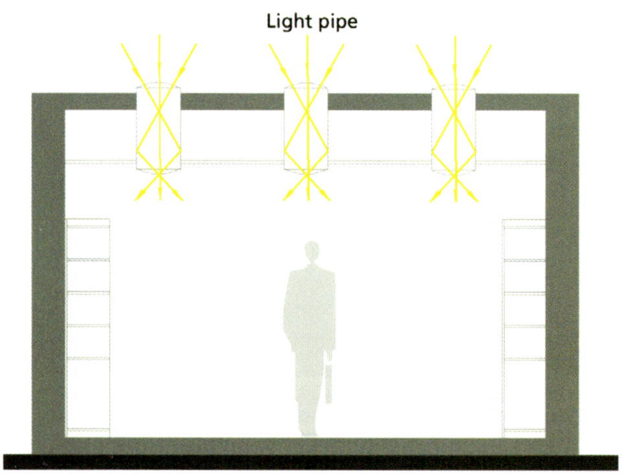

(a)

Figure 28: Daylight guiding systems **(a)** light pipes **(b)** mirror shafts

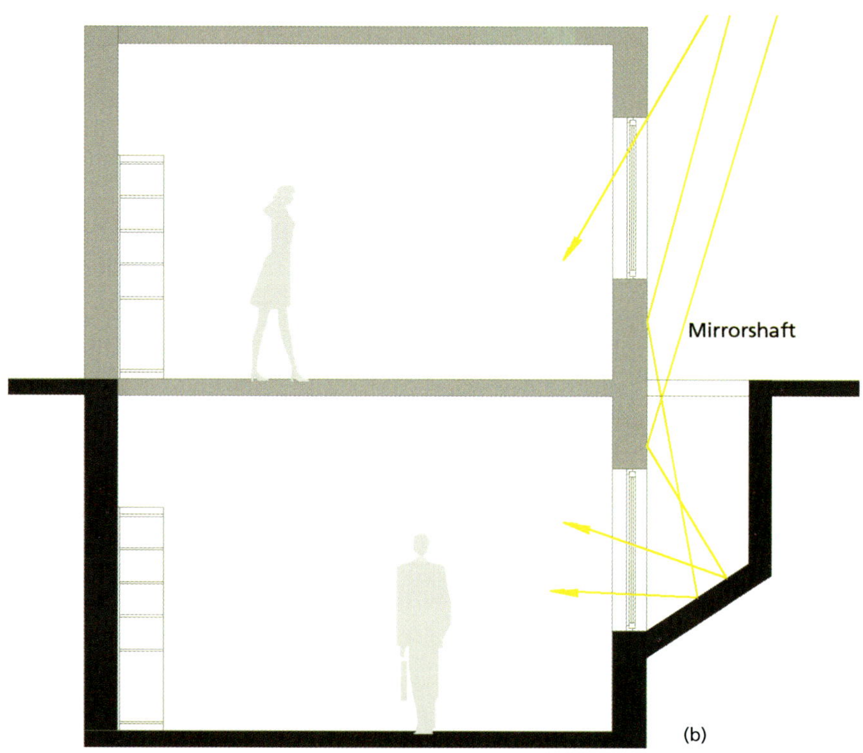

(b)

CASE STUDY Sainsbury's, Dartmouth

Sixty-four 750 mm diameter light pipes are installed in the main shop floor area, with additional units used in the offices and other areas of the store (Figure 29). Automatic lighting controls are in place to dim electric lighting when enough daylight is present in the store areas, and occupancy detection is coupled with daylight-based controls in staff areas and toilets. Using light pipes for daylighting and automatic lighting controls leads to 50% energy saving and 40% reduction in carbon emission, helping the store achieve a BREEAM 'Excellent' rating[23].

Figure 29: Light pipes used in the sales area, combined with daylight-based dimmable electric lighting, help reduce the energy use significantly (© Monodraught)

Figure 30: Automatic dimming of electric lighting compensating for daylight variability (© Philips Lighting)

Optical systems are able to redirect or harvest daylight by means of tubular light-guiding systems, fibre optics, or arrays of mirrors and lenses. Daylight-guiding systems (Figure 28) can lead daylight collected at roof level into internal spaces via highly reflective, mirrored tubes, and are more effective under clear sky conditions, although some of them can deal quite effectively with diffuse skylight. Fibre optic systems rely mainly on direct sunlight, as they are typically integrated within complex arrangements using sun-tracking collectors. Sophisticated arrangements of mirrors and lenses can be used to direct daylight (mainly direct sunlight) into specific internal areas of interest[22].

4.4.2 Integrating daylight with electric lighting

Electric lighting should be designed to function and be controlled in conjunction with the daylighting strategies. Appropriate attention should be paid to the choice of lamps, luminaires and controls, as well as to defining the correct control zones, based on daylight penetration into internal spaces. An effective control system ensures that the electric lighting operates only when needed, depending on the type of space, reducing wasted energy and improving building energy performance and profitability. Automatic dimming that compensates for daylight variability is the most effective and flexible approach to controlling electric lighting in daylit spaces (Figure 30).

Significant energy savings can be achieved by correct zoning of the electric lighting – that is, by establishing the groups of luminaires that should be controlled simultaneously, based on different factors that include daylight availability. Typically, luminaires are grouped together in each zone that receives a similar amount of daylight.

4.5 Intelligent lighting controls

Appropriate lighting controls[24] form an essential part of the lighting system. They can save energy when properly specified, up to 30–40% or more in some types of building. There is a wide variety of control types, ranging from simple switches to complex integrated lighting control and dimming systems.

4.5.1 Manual switches and dimmers

In retail environments, centralised manual control is usually carried out from control panels to switch or dim selected different parts of the lighting systems, or to enable pre-programmed scene setting. Scene setting allows the store manager to select different patterns of lighting – for example a 'daytime' setting and a warmer, dimmer, 'evening' setting. Alternatively, 'scenes' with reduced lighting provision can be used during cleaning and restocking. The more sophisticated control systems should allow the store manager to make changes to the way lighting is switched or dimmed, rather than calling in the lighting control supplier to make changes every time lighting is altered, or a different pattern of control is required. Whatever the type of control, it is a good idea to have a single, easy-to-identify 'OFF' switch that the last person to leave the store can use to turn out all the lighting.

Manual switches lead to energy waste if lighting is left on when not needed. Often, store managers are too busy to make constant changes to the lighting, so if lighting needs to respond to changing daylight levels and occupancy, automatic control is the best option.

4.5.2 Presence and absence detectors

Presence and absence detectors generally use passive infrared sensors to detect whether people are in the scanned area, but can also be based on ultrasonic or microwave technology. Microwave sensors are better for large areas. Presence detectors turn lighting on when they detect a movement, and off after a preset time with no movement detected. They are ideal for stockrooms, where people entering may have their hands full, and not have time to search for light switches. Placement of sensors needs care, as a sensor may not detect someone behind shelves full of stock, and may not respond to someone walking directly towards it. Normally a sensor should be placed so that someone entering the space moves across its field of view. If people are likely to spend some time working in the space, additional sensors in different areas may be needed.

Full presence detection can result in unnecessary energy use if lighting is switched on when it is not required, for example in daylit areas. In these circumstances absence detection is best, as the occupancy sensor only switches lighting off, and switching on is by manual control. All detectors must be compatible with the lamp type. Very frequent switching can reduce the life of fluorescent or metal halide lamps.

4.5.3 Daylight sensors

Automatic daylight sensors (photocell controls) can switch off or dim lights when there is sufficient daylight to provide adequate lighting. They are often used for external lighting, but can also be integrated in control strategies for internal lighting where they are mounted on the ceiling in zones with daylight. A time delay is usually built in to cater for short-term variations in daylight caused by, for example, a passing cloud. Photocell controls can be integral to fixture, switch or stand-alone, in which latter case their compatibility with fluorescent, metal halide and LED lamps must be checked. Daylight-based dimming of metal halide lamps is currently limited to some lamps, and typically down to 50% only.

4.5.4 Timers

Timers can be preset to switch lighting on or off, or dim it, at predetermined times. They may be used for:

- Switching or dimming lamps out of main trading hours to a lower level, to allow for cleaning and restocking. For example, one third of the lighting may be left on. Sophisticated systems can leave a different third on each night, to give an even lamp life across the whole installation. In major stores this approach has been shown to save 20% of total energy use.

- This particularly includes lighting in chiller cabinets. They often contain a 55 W fluorescent lamp, which then generates another 30 W or so in cooling load, as the heat from the lamp has to be removed. They are often wired on a single electrical supply, so that the lamp is left on all the time the cabinet is switched on, even at night. There are considerable, and highly cost-effective, savings from switching this lighting off outside trading hours.

- Out-of-hours switching to light display windows only.

- Switching off window display lighting (and exterior lighting) when there is no longer significant pedestrian traffic outside, for example in the late evening.

Time delay switches are typically push-to-operate switches, often used in corridors, that automatically switch lights off after a preset time. In safety-critical areas such as staircases there is a danger that the lighting can go off when someone is still in the space; presence detectors are usually safer. Solar time switches are useful for controlling external lighting, as they adjust the on/off times to coincide with dawn and dusk throughout the year.

4.5.5 Management systems

There are various energy management/lighting control systems that range from the relatively simple, which can switch lamps off by momentarily interrupting the power supply, such as electronic key fobs, to those that operate through sophisticated signalling systems, as with modern automatic dimming controls that can be programmed to change the lighting levels gradually and indiscernibly to produce the desired ambience at preset times. A building management system (BMS) can switch lighting at pre-programmed times, or operate complex lighting systems via multi-channel switching with different switching programmes for each channel.

© iGuzzini: Napapijri, Milan, Italy

5 Guidelines for improved lighting design

There are no standard solutions for lighting sales areas that fit a specific type of store or merchandise. Custom solutions need to be chosen, based on the characteristics of the space and type of merchandise. The following guidelines aim to cover most of the typical situations that are met in retail spaces.

Retail lighting is fundamentally a combination of general, accent, display, and architectural and decorative lighting. The design scheme adopted depends on the store profile and the associated factors described in Section 2.2.

An improved retail lighting design aims to:

- emphasise the store's character
- create a pleasant atmosphere for shoppers
- enable easy and accurate examination of the features and qualities of the merchandise
- minimise glare and visual discomfort
- be energy-efficient.

Typical strategy for retail lighting design

- Review the type of store, and the functional, economical and aesthetic design factors, such as space appearance and layout, lighting technologies of interest, and correlation or interference with other building services.
- Assess daylight penetration, and the potential for improvement.
- Review targets for lighting design: illuminance levels, colour temperature, colour rendering.
- Establish the balance of general, accent and display lighting.
- Choose static or dynamic lighting.
- Identify the areas to be accented.
- Decide whether walls or space need special lighting.
- Choose lamp types and parameters.
- Choose luminaires for each lighting arrangement.
- Perform quantitative and qualitative analysis of the chosen lighting system; make necessary changes until targets are met.
- Choose appropriate lighting controls in each area.

Lighting design varies with the functional requirements of the lit spaces. In general, a retail environment consists of four main types of space: sales areas, display windows, ancillary spaces, and exterior spaces. Depending on the store profile and target market, the distribution and direction of light need to be defined in the initial design stages for each type of space.

5.1 Sales areas

The lighting distribution in sales areas can range from uniform, diffuse light to a high-contrast scheme using focused light. In most cases a combination of the two is used.

5.1.1 General lighting

General lighting in the sales area should aim to permit easy navigation through the main aisles and examination of merchandise, while emphasising the store's image, and creating a desirable, visually comfortable ambience in the store. It should produce a glare-free, uniform illuminance throughout the store. According to the store profile, the maintained average illuminances typically range from low light levels (50–300 lx) in high-end, exclusive stores to high values (500–1000 lx) in mass-merchandising stores, with average levels (300–500 lx) in middle-category stores (ie department stores).

The overall recommendation of the European standard BS EN 12464-1:2011[10] for retail premises is an average general lighting illuminance of 300 lx for sales areas and 500 lx for till and wrapper areas. In order to improve energy efficiency, lower light levels similar to those recommended in BS EN 12464-1:2011 could be used, especially in mass-merchandising stores.

To draw attention to the merchandise, it is better to create contrasts and highlights rather than just have an overall bright light level in the entire sales area. This involves lower general lighting levels, combined with accent lighting to produce contrasts and bright light to highlight the merchandise. The retail space is enhanced by the individual displays, and becomes more inviting as the ambient lighting is reduced. The appearance of the space, and shoppers' sense of well-being, will also be improved if diffuse general lighting is supplemented by vertical lighting that delimits the periphery of a space[25].

Colour temperature and colour-rendering qualities vary with store profile. A neutral to cooler white colour appearance (3500–5300 K) tends to be more common in low-end, mass-merchandising stores, and a neutral white colour appearance (3500–4000 K) is generally used in middle-category stores (ie department stores), whereas the colour appearance in high-end, exclusive stores is typically warm (2700–3000 K). Colour rendering is usually above 80 in all cases, with higher values (above 90) in exclusive stores.

Research[25] has found that the most appropriate colour appearance for the general lighting system is neutral white, as it makes shoppers spend more time in a store, and improves their sense of well-being. Warm white light creates a sense of familiarity, safety and security, whereas cool white light creates an impression of more spacious areas. Varying colour temperatures within a lit space can provide a dynamic character that is more attractive than a scheme using the same colour appearance for all light sources. LED lighting using advanced colour-tuneable technology can vary the colour temperature of the light emitted. However, uncomfortable colour contrasts should be avoided. In adjacent areas, colour temperatures should not vary too much (not more than 800 K to 1000 K difference).

Traditional design schemes comprise static installations that provide constant illuminances throughout typical retail spaces. Apart from signage lighting used for advertising, which attracts observers' attention by the combined use of colour, motion and size, all other lighting typically provides a static ambience. However, recent research on the behaviour of supermarket customers carried out by Philips Lighting and the Plus supermarket group showed that dynamic retail lighting results in more sales than traditional static installations. Through the use of different lighting settings for colour temperature and light levels, applied over alternating days of the week, combined

CASE STUDY M&S Simply Food, Sheffield

Store specifications required that the lighting load be reduced from 11 W/m² to 7 W/m² (Figure 31). The lighting design approach reduced ceiling fitting height in the aisles from 3 m down to 2.8 m, concentrated light levels on the vertical surfaces of the merchandise, and reduced ambient levels to 300 lx in some places.

An all-LED lighting installation[26] (from the store to back-of-house, and from chiller rooms to the car park) was supplemented by light pipes for daylighting. Fittings used were twin-panel with a 'batwing' distribution (panels mounted at an angle of 30°). LED spotlights created sparkle. Automatic controls provided daylight-based dimming and information on energy consumption.

The scheme is 25–30% more efficient than the lighting systems used in a typical M&S Simply Food. The store was about 6% more expensive to build than a standard M&S Simply Food, but achieved a BREEAM rating of Excellent by incorporating a whole range of sustainability measures (FSC-certified timber, reclaimed bricks, rainwater harvesting, heat recovery from refrigeration units to help heating and hot water, green roof and green wall encouraging wildlife, recyclable materials). Carbon emissions were cut by 23% and energy use reduced by 30% compared with a conventional store of similar size.

Figure 31: LED lighting can help reduce the installed wattage and energy consumption (© Philips Lighting)

with a dynamic LED wall (Figure 32), the average sales per customer increased by 1.9%[27]. Furthermore, it was found that customers tended to spend more time in the areas with warmer light settings than in those with cooler ones. On the other hand, the same case study suggests that when the store is quiet, warm lighting can ensure that people stay longer in the store, whereas when the store is busy, cool lighting helps them make decisions faster, thus speeding up the flow of customers[28].

General lighting is commonly achieved through a combination of three basic approaches[3]. The general pattern system and the specific pattern system, both based on a fixed fitting layout, and the flexible pattern system. The choice of these approaches needs to be based on the architectural interior design, and on the flexibility required for the selling process:

- The general pattern system consists of a regular array of luminaires that provide general lighting in the sales area, without taking the location of the merchandise into consideration.
- The specific pattern system comprises a layout of luminaires that takes the location of the merchandise into account, and is tailored to emphasise the merchandise and delineate regions of the sales area.
- The flexible pattern system allows temporary installation of luminaires in changeable arrangements that can be used for general or specific pattern lighting that is tailored to the merchandise display.

Linear fittings can be positioned parallel to racking or perpendicular to racking. Having runs of lamps perpendicular to the racking allows for future flexibility, in case there are changes in rack positioning later on. However, from a lighting point of view, the better solution is to have fittings parallel to the racking. This allows luminaires to be chosen with a controlled optical distribution to direct light effectively towards the merchandise.

Figure 32: An LED light wall can display a colour-changing logo to enhance brand identity and support communication and connection with shoppers (© Philips Lighting)

The appearance of the luminaires used needs to be consistent with the store profile and style. Fixed pattern systems generally use arrays of luminaires that range from bare fluorescent lamp battens or recessed fluorescent fittings in low-budget stores, through to luminaires integrated with the store architecture, such as cove lighting or suspended uplights, in stores with quality or exclusive profiles. The flexible pattern typically employs recessed, adjustable luminaires and track-mounted luminaires.

The most efficacious lamps recommended for general lighting in sales areas are triphosphor fluorescent lamps (T5 for enclosed fittings), metal halide lamps (including high-bay spaces or uplighting in certain applications), or the newest generations of LED lighting. High-frequency electronic ballasts and LED drivers with a higher power factor are the most efficient, and are recommended.

5.1.2 Accent lighting

In addition to general lighting, accent lighting is used in specific areas to emphasise particular features of the merchandise (shape, texture, finish or colour), and to provide an expressive variation in brightness and shadow throughout the store. Apart from mass-merchandising stores, retailers use some level of accent lighting to attract shoppers to the products displayed. Visibility is improved and a stronger visual impact is created, drawing shoppers out of the main aisle into the merchandising space[3].

Wall-washing schemes increase the brightness of vertical surfaces (ie walls and display units), making a store environment look larger, more pleasant, and more attractive. Linear or point light sources can be integrated using various techniques to provide continuous or individual light patterns.

Highly efficient fluorescent or linear LED wall-washing fittings are generally the most energy-efficient option for large area wall displays, where small-area accent lighting can be best effected by compact metal halide or LED adjustable spotlights (Figure 34). By adopting track-mounted or recessed adjustable flexible arrangements, the accent lighting can be easily adapted to suit various layouts and types of merchandise.

In all cases, precise beam control is required, and the vertical surfaces should be illuminated as uniformly as possible from the top to the bottom. Accent lighting requires brightness higher than the surround to highlight the merchandise and displays; hence accent illuminances can be as high as 10–15 times of that in the surround. Special consideration should be given to placing and orienting the light sources in such a manner that glare from direct or reflected light is avoided for both customers and store staff.

5.1.3 Display lighting

Display lighting aims to attract customers' attention to merchandise open to examination, or to merchandise in showcases, by a combination of brightness, colour and

CASE STUDY Mulberry, London

The store received a BREEAM rating of Excellent, and consumes 50% less energy than any comparable store in the chain.

Ceiling light boxes simulate daylight by a combination of warm and cold white fluorescent tubes (Figure 33).

Integrated LED Light Sheet panels are fitted to the underside of shelves to resemble the daylight-simulating light boxes. Highly efficient 35 W metal halide track and spot fittings provide flexible accent lighting[29].

Figure 33: Shelf-integrated LED panels resembling daylight-simulating fluorescent light boxes (© LED Light Sheet)

Figure 34: Adjustable spotlights can be aimed at specific small areas of interest to create accents (© Megaman)

modelling. High brightness contrasts attract more attention, and the appropriate ratio of luminance between the object lit and its background can be chosen according to the effect desired. Table 4 shows the strength of accenting for various luminance ratios.

Different effects can be obtained by using chromatic schemes with saturated colours. 'Strongly coloured light on an object of the same colour will deepen the colour, whereas strongly coloured light on the background and surroundings will change the atmosphere'[5].

Modelling can emphasise the forms and textures of the merchandise by creating luminance contrasts through the appropriate use of directional lighting. The modelling achieved depends on the relative strength of light delivered from different directions, by a combination of key-light, fill-light, backlight and uplight techniques, described in Table 5[5].

Furthermore, the features of the merchandise are best revealed if the most appropriate modelling technique is used for a specific material. Table 6 presents some of the more common display lighting techniques for different materials.

Eye-tracking studies[25] showed that the lower levels of shelves go relatively unnoticed by clients, and the use of accent or dynamic lighting in the lower third of shelves can favour sales. At the same time, wide-area backlighting of shelves creates more attractiveness than accent lighting only. Hence a combination of backlighting and accent lighting seems to have the best effects on the merchandise. Point, linear or surface light sources can be used with LED panels, proving to be an ideal

Table 4: The influence of luminance ratio on the strength of accent lighting[5]

Luminance ratio (accent/background)	Strength of accenting
1	None
2	Noticeable
5	Low theatrical
15	Theatrical
30	Dramatic
> 50	Very dramatic

Table 5: Components of display lighting[5]

Light	Description	Function
Key-light	The main source of directional lighting	To create sparkle and reveal texture
Fill-light	Supplementary lighting from a different direction	To soften shadows so as to get the contrasts in the display at the desired level
Backlight	Lighting from behind and usually above	To separate the object from its background; to reveal transparent elements
Uplight	Lighting accentuating parts of the display close to the floor	To soften shadows; can be used for dramatic effects

Table 6: Common display lighting techniques for particular materials[5]

Materials	Display lighting technique
Uniformly transparent materials	Transmitted light from a lit background; uplighting, possibly in colour
Glass and crystal	Highlighting; uplighting, possibly in combination with translucent background lighting; coloured light
Transparent fibrous objects, eg fine textiles	Contour lighting from behind
Precious stones and jewellery	Small spotlights, black velvet background
Opaque, shiny objects, eg silver	Spotlights, black velvet background, highlighting
Opaque, textured objects	Light predominantly glancing across the surface

option for backlighting by size, capacity to match any colour temperature or colour-changing scheme, low heat emission (very important for heat-sensitive products), and long life. Being most efficient at lower temperatures, LEDs are ideal for display lighting of refrigerated spaces.

Shelf lighting (Figure 35) is typically used in high-end stores to attract shoppers and enable easy evaluation of the products, which need to be fully lit and their colour and texture realistically rendered. For clothing, light sources with similar colour rendering need to be used for shelf lighting and in fitting rooms. Concealed light sources, such as CDM, linear fluorescent or LEDs, can be used for shelf lighting, and aiming needs to be carefully assessed to avoid glare.

An effective approach to shelf lighting consists of creating an attractive wash of light from above, evenly distributed from the top to the bottom (Figure 36). Often this can be achieved by combining dedicated wall-washing fittings over the shelving with supplementary directional lights such as track-mounted spotlights in the main sales area[30]. The fittings need to be positioned at the right distance from the shelf edge and at the right angle to create the uniform light distribution across the entire vertical shelf surface.

Figure 35: Various arrangements of shelf lighting (© acdc)

Display cases can incorporate special lighting to make merchandise more attractive to shoppers in high-end stores. Backlighting or linear arrangements can provide a uniform level throughout the display case, with point sources adding sparkle, especially for items such as glassware or jewellery. Linear fluorescent and LED lamps offer good efficacy and flexibility, with cold cathode tubes and fibre optics allowing for various shapes that might be required by the display case design. For display cases with glass panels, ambient light reflected off the glass can affect merchandise visibility, but this can be solved by higher illuminance in the display case compared with ambient light levels, or by carefully aiming the other light fittings to avoid light being reflected at the shoppers' eye level.

Chiller cabinets can use display lighting to improve merchandise visibility, particularly in sales areas with relatively high ceiling height, and for vertical cabinets. Although the most common solution is currently to use linear fluorescent lighting, this is not the most energy-efficient, as the efficacy of fluorescent lamps is highest at 25°C (T8 tubes) or 35°C (T5 tubes). Linear arrangements of LED lamps, correctly aimed at the merchandise and concealed from the shoppers' view, are recommended, given the higher luminous efficacy of LEDs at lower temperatures.

Figure 36: Attractive wash of light, evenly distributed from the top to the bottom (© Erco)

5.1.4 Decorative and signage lighting

General lighting of sales areas can be supplemented by architectural and decorative lighting to enhance ambience and set particular moods in specialty stores or high-end department stores. Architectural cornices can conceal cove lighting using T5 fluorescent seamless fittings, cold cathode fittings, or LED colour-changing linear fittings. Decorative lighting schemes can incorporate chandeliers, floor-standing fittings, wall-, ceiling- or ground-mounted luminaires, or light graphics. Recent developments in LED technology and lighting controls provide various efficient and flexible solutions for decorative lighting. It is essential that decorative lighting does not detract from the lighting scheme aimed at the merchandise.

Stairwells offer an opportunity for both architectural and lighting techniques to attract and guide shoppers to areas of interest in the store. Spotlights can highlight architectural features, or draw attention to complex construction details. An escalator can be transformed from a functional transport system into an attractive feature by carefully combining light reflections and transparency effects with correct light fittings to animate the setting, and maintain customers' curiosity about the goods yet to be seen (Figure 38).

Illuminated signs add advertising value to any store, attracting the attention of passers-by through static or dynamic combinations of shape, size and colour.

CASE STUDY John Lewis at Home and Waitrose, Ipswich

Figure 37: (a) John Lewis at Home **(b)** Waitrose. LED light sources used for accent and display lighting in shelves and display cabinets

This is the first destination to have a John Lewis at Home store and a Waitrose supermarket on the same site (Figure 37). Opened in November 2012, the branch was fitted out entirely with LED lighting (sales areas, office spaces and warehouses), which helped reduce the lighting load to below 10 W/m^2.

John Lewis at Home
Stepping away from the traditional CDM solution, John Lewis at Home is lit by LED spotlights and floodlights aimed at the merchandise that are fitted with 3000 K warm white and CRI 88 LED chips. The fittings can be re-aimed when changing merchandise position across the sales space. The lighting scheme, used primarily for accent and display lighting, was designed to create contrasts and highlights that would be visually appealing to customers. The planned replacement of the current second-generation GE Lighting Infusion LED modules with third-generation Infusion chips will lead to a further reduction of the lighting load to 7 W/m^2.

Waitrose
The sales area in Waitrose is lit by a combination of ceiling-recessed square modules and round spotlights using LEDs with a colour temperature of 4000 K, while chiller cabinets are lit by 4200 K CRI 80 LED tubes concealed from the shoppers' view and aimed at the merchandise. This delivered a well lit solution at less than 9 W/m^2 and achieved a further energy reduction over Bracknell (see case study on page 7) of approximately 40%.

Figure 38: Using light reflections and transparency effects can transform escalators into attractive features that invite customers to explore the shopping centre (© Philips Lighting)

Luminous letter signs using illuminated characters on a darker background typically incorporate exposed lamps or exposed luminous tubes. Although incandescent lamps in either clear and coloured bulbs or gas-filled tubes were traditionally used, LEDs are currently the most suitable lamp type for such applications, given their small size, control possibilities including high-speed dynamic effects, and better performance in an open environment. Fibre optics can also be used for sign backlighting, decorative and edge lighting, or lettering (side-emitting fibres), or to add sparkle or colour to static signs (end-emitting fibres).

Luminous panel signs are illuminated by internal light sources, and use a translucent plastic or glass cover to convey the advertising message. Fluorescent, metal halide and LED lamps are recommended for luminous panels, with appropriate heat sinks in place for LEDs to ensure energy-efficient performance. For coloured signs, LEDs of the appropriate colour are the most energy-efficient option.

Floodlit signs are typically used outdoors for advertising, and require sufficient brightness to stand out from their surroundings. Metal halide is an energy-efficient choice for sign floodlighting, and linear fluorescent or linear LED arrangements can also be used on the top and/or bottom edge of the sign. The location of the light sources needs to be carefully chosen in order to avoid uncomfortable reflected glare or shadows. Shielding might also be required to avoid visual disturbance from lamps to traffic facing the direction of the sign. Careful aiming is needed to ensure uniformity, and to avoid spill light.

External signs need to be impact resistant and appropriately protected against ingress of dust and moisture.

Recently developed LED screens can also be used for store advertising. Such devices are outside the scope of this guide.

5.2 Display windows

The function of display windows is to attract the attention of passers-by and make the merchandise displayed look appealing. Therefore display window lighting needs to be brighter than its surroundings, and minimise any reflection that could impact on the appropriate visibility of the displayed items. According to the *IESNA Lighting Handbook*[3], three times the illuminance of the circulation area lighting is required for showcase lighting in shopping centres.

Lighting of display windows needs to take into consideration daylight that is reflected from the window glass, and from the windows of nearby stores[5]. Display windows generally incorporate accent lighting on merchandise, in addition to daylight. Depending on the amount of daylight and the lighting of adjacent stores, the general lighting of display windows during the day needs to be in the range 500–2000 lx, with accent lighting increasing illuminance to values in the range 3000–10,000 lx. Values towards the higher end are typically chosen to light up items of lower reflectance (around 0.2–0.3) in order to render an appropriate luminance. The lighting levels should be decreased after dark and when there is little daylight, as the surround is significantly less bright, and contrasts can still be achieved with lower light levels and hence lower energy consumption. Dimming and multi-step on/off controls for the electric lighting can compensate for daylight variability, and provide facilities for lower light levels for night-time conditions[30].

A typical lighting system for display windows consists of three main components: general lighting that lights up the display window uniformly; accent lighting that provides modelling and highlighting; and backlighting. The principles described previously for display lighting in sales areas typically apply for display windows as well. Lighting at an angle provides a better effect than downlighting on its own[30]. The merchandise is emphasised by creating contrasts with its surrounding area (Figure 39). It is more energy-efficient to create high contrasts if lower general lighting levels are chosen. However, the light levels for general and accent lighting need to be chosen according to the daylighting conditions available at various times of day, to make sure that the merchandise is always correctly highlighted.

In general, providing highlighting during the daytime can be inefficient, owing to the high light levels of electric lighting required, and their associated electricity consumption. In such cases, electric lighting needs to make display windows brighter than the daylit surroundings nearby, and to diminish the effects of daylight reflection from the window glass. A solution to avoid daylight reflection that also enables the use of lower general lighting levels for display windows would be to alter the reflection angles by changing the verticality of the window glass. An outward-tilted window glass is recommended when the ground in front of the display window is not lit by direct sunlight, or when its reflectance is low. An inward-tilted window glass can be used in combination with a low-reflectance overhang mounted above the display window. Having an overhang or canopy above the window provides shelter for shoppers, as well as reducing reflection from the glass.

The most energy-efficient light sources currently available for display windows are, depending on the space characteristics and the lighting effects intended, compact metal halide lamps, point, linear or surface LEDs, or T5 fluorescent lamps for backlighting. Colour-changing effects can be best put into practice by the use of fully controllable LEDs.

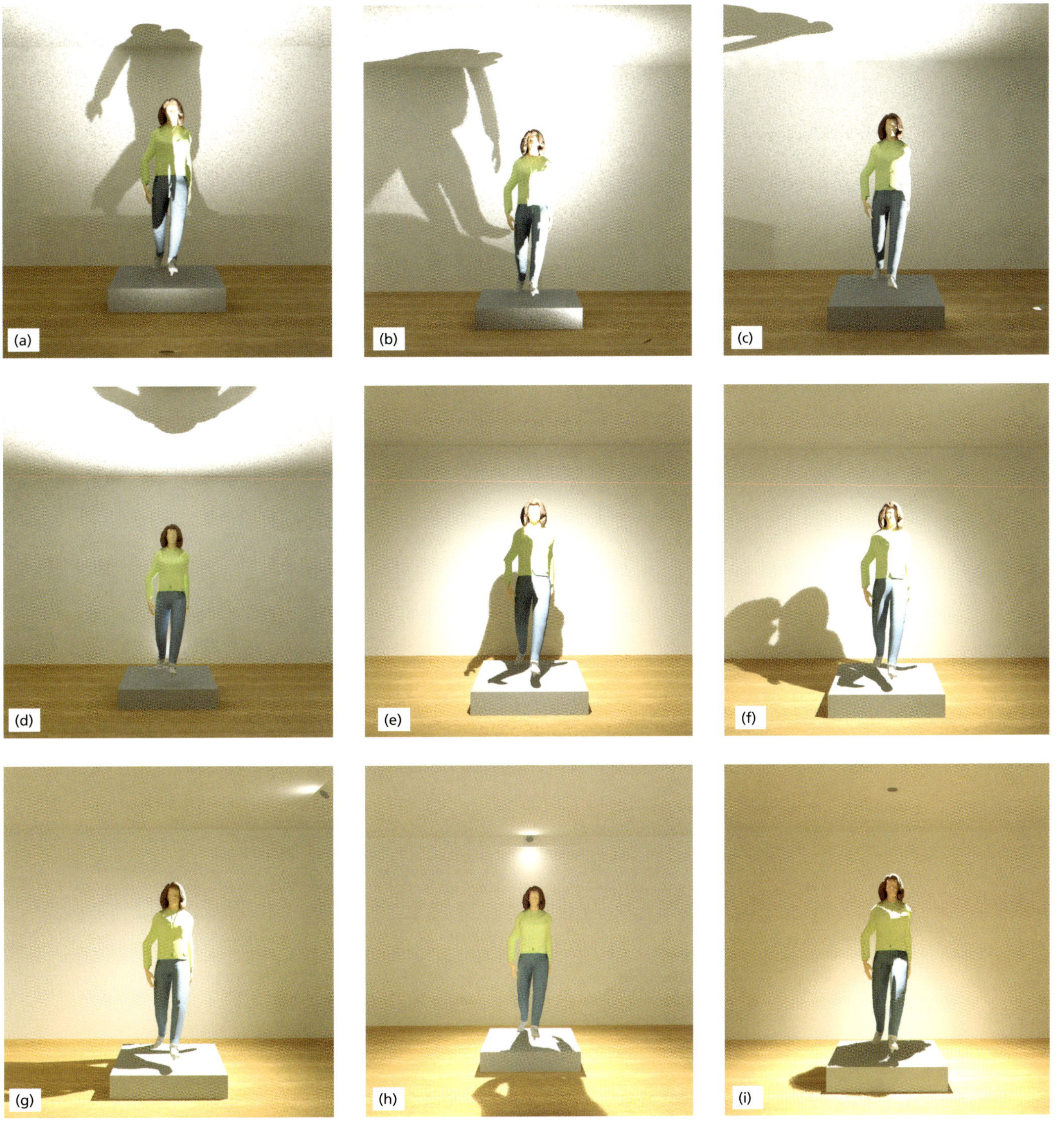

Figure 39: Various light effects can be achieved on mannequins in display windows by placing the light fitting at various positions: **(a)** bottom-mounted, front **(b)** bottom-mounted, 45° left **(c)** bottom-mounted, left **(d)** bottom-mounted, behind **(e)** top-mounted, front **(f)** top-mounted, 45° left **(g)** top-mounted, left **(h)** top-mounted, behind **(i)** top-mounted, overhead

5.3 Ancillary spaces

The ancillary spaces of a retail area include fitting rooms, alteration workrooms, wrapping and packing areas, lounges, stairwells, corridors, public toilets, and indoor car parking. Lighting recommendations for these areas are given in this section.

5.3.1 Fitting rooms

Lighting of fitting rooms requires crucial attention in clothes stores, as this is often where the customers decide whether to buy. The layout and facilities available in the dressing and fitting rooms depend on the store profile. However, in all cases it is essential to provide ideal visibility by ensuring appropriate vertical illuminance, very good colour rendering, elimination of harsh shadows, correct modelling of faces and bodies, and suitable colour appearance. Lighting schemes are typically based on a combination of diffuse and directional lighting solutions.

Overhead lighting creates modelling effects and enhances texture, but needs to be carefully placed to provide optimal vertical illumination down to the floor. Ceiling luminaires tend to create shadows, and an undesirable view for customers. Hence wide-distribution downlights and ceiling fittings are preferred.

Diffuse mirror lighting should supplement overhead lighting to soften shadows without reflected glare. The light sources need to be positioned so that the customers standing in front of the mirror are lit, rather than the mirror itself. Energy-efficient T5 fluorescent or linear LED fittings can be placed behind mirrors to create a halo effect.

A study carried out in the US[31] investigated how the direction of lighting in a dressing room influences the shopper's emotional state, perception of lighting, preference, self-evaluation and whole dressing-room experience. The results indicate shoppers' preference for frontal lighting (consisting of diffuse mirror lighting) rather than overhead lighting (done by track-mounted spotlights), and also the importance of lighting in evaluating the overall dressing-room experience: 51% of the comments people made were about lighting, compared with 49% for all the other design factors.

A flexible combination of overhead and sconce lighting can be incorporated in variable light settings that simulate 'day', 'dusk' and 'night' to allow customers to evaluate the merchandise under the desired light scenario (Figure 40), which can be chosen manually from wall-mounted controls[32].

Light sources with high colour rendering (above 90) and warm appearance (2700–3300 K) are optimal for lighting dressing and fitting rooms. To encourage energy efficiency, compact metal halide, LED or compact fluorescent lamps are the best options for overhead lighting, with T5 fluorescent or linear LED lamps a recommended option for mirror lighting.

Energy savings can also be achieved by using optimised control schemes that include presence-based dimming or switching. The retailer's preference may be to maintain lighting levels at a reduced level when individual fitting rooms are unoccupied, rather than switch off the lights in those spaces. Overhead and/or mirror lighting can come on when presence is detected, either fully or, for a softer appearance, gradually to full load.

Figure 40: Flexible controls allowing shoppers to change lighting settings in the fitting rooms, between simulated 'day', 'dusk' and 'night' (© Boux Avenue)

5.3.2 Other spaces

Other areas in store environments where customers typically have access include alteration workrooms, wrapping and packing areas, lounges, stairwells, corridors, public toilets, and indoor car parking. Guidance on lighting these general areas is given in the *SLL Lighting Handbook*[5].

Bank counters can be considered as workplaces, but there is also an interaction with customers. From this perspective, appropriate lighting levels need to be provided for customers and banking personnel to see each other properly, but also on the counter so that customers can undertake reading or writing tasks comfortably. In both cases, glare from direct light coming from luminaires and from light reflected off counter glass, as well as harsh shadows, needs to be avoided by correct aiming of the light beam. Fittings with a wide beam angle can help meet these requirements.

Figure 41: The store's character and architectural elements can be emphasised by attractive architectural lighting strategies (© Philips Lighting)

5.4 External spaces

Two types of lighting are typically used in external spaces: lighting for access and parking facilities, and architectural lighting. These guidelines do not address the first category, for which extensive guidelines are provided by appropriate standards[33]. Architectural lighting should incorporate energy-efficient solutions using compact metal halide, LED or fluorescent lamps as point, linear or surface light sources. Although less energy-efficient, colour-changing schemes, typically using LEDs, provide a dynamic, vibrant and appealing character to the store environment.

External lighting should emphasise the store's character and image, highlighting entrances and architectural elements (Figure 41). A well-thought-out strategy should be conceived to attract and guide shoppers to the sales areas.

General guidelines for external lighting in retail applications should take into account specific factors, such as[3], [33]:

- allowing easy identification of key areas, ie entrances, exits, parking facilities, circulation areas
- enabling safe routes for driver and pedestrian access
- contributing to the security of people and the site
- providing visual unity for the shopping area
- reproducing the image and ambience of the retail establishment.

© Philips Lighting

6 Conclusions

Lighting constitutes a major energy use in retail stores. A survey carried out by BRE revealed that the average lighting power density was 36 W/m^2, compared to 10–12 W/m^2 in a modern office with efficient lighting. The same survey found that the majority of retailers relied on traditional lighting solutions. Three quarters of lamps were fluorescent, and nearly half the remainder were inefficient, short-lived tungsten halogen lamps.

The case studies in this guide show that it is possible to have energy-efficient lighting solutions that enhance the retail environment and encourage sales. Key strategies include:

- reduced general lighting levels with accent lighting to highlight displays and wall lighting, to make the space look bright and appealing
- efficient luminaires with high light output ratio to maximise lighting provision
- the use of energy-efficient LEDs and compact metal halide lamps for spotlighting and accent lighting, rather than tungsten halogen
- careful aiming of spotlighting and floodlighting to maximise the impact of displays and reduce glare to staff and shoppers
- the controlled admission of daylight
- lighting controls to make the most of daylight and reduce lighting levels outside trading hours
- LED lighting in chiller and freezer cabinets, with controls to turn this off when the store is shut.

As well as improving the appearance of the store, such energy-efficient lighting solutions can reduce maintenance costs as lamps last longer, and can reduce overheating and cooling costs.

© Philips Lighting

7 References

1 Department of Energy & Climate Change (DECC). Energy consumption in the UK. London, DECC, updated July 2012. Available online at: www.decc.gov.uk/en/content/cms/statistics/publications/ecuk/ecuk.aspx.

2 Building Research Energy Conservation Support Unit (BRECSU). Energy efficient lighting in the retail sector. Good Practice Guide 210. Watford, BRE, 1997.

3 Rea M S (Ed). The IESNA lighting handbook: Reference & application. New York, Illuminating Engineering Society of North America, 2000, 9th edn.

4 Custers P J M, de Kort Y A W, Ijsselsteijn W A and Kruiff M E. Lighting in retail environments: atmosphere perception in the real world. Lighting Research & Technology, 2010, 42 (3) 331–343.

5 Society of Lighting and Lighting (SLL). The SLL lighting handbook. London, Chartered Institution of Building Services Engineers, 2009.

6 Quartier K, Vanrie J and van Cleempoel K. Atmospheric lighting in supermarkets. Proceedings of the 2nd CIE Expert Symposium on Appearance. Gent, Belgium, 8–10 September 2010.

7 Schielke T. Light and corporate identity: using lighting for corporate communication. Lighting Research & Technology, 2010, 42 (3) 285–295.

8 Faithfull M. Waitrose targets BREEAM at Bracknell. Lux Magazine, January 2012, 10–16.

9 See http://etl.decc.gov.uk.

10 BSI. Light and lighting. Lighting of work places. Indoor work places. BS EN 12464-1:2011. London, BSI, 2011.

11 Loe D L, Mansfield K P and Rowlands E. Appearance of lit environment and its relevance in lighting design: experimental study. Lighting Research & Technology, 1994, 26 (3) 119–133.

12 Heschong Mahone Group (HMG). Skylighting and retail sales. An investigation into the relationship between daylight and human performance. Detailed Report for Pacific Gas and Electric Company. Fair Oaks, CA, HMG, 1999.

13 Tilley M. Embrace change or pay the price. Lux Magazine, April 2011, 32–33.

14 Talking shop – Forum Retail. Lux Magazine, February 2011, 50–58.

15 Vieira M. Within limits. Lux Magazine, October 2011, 66.

16 Molony M. An LED future in store. Lux Magazine, January 2011, 56–58.

17 European Commission. Commission Regulation (EC) No 245/2009 of 18 March 2009 implementing Directive 2005/32/EC of the European Parliament and of the Council with regard to ecodesign requirements for fluorescent lamps without integrated ballast, for high intensity discharge lamps, and for ballasts and luminaires able to operate such lamps, and repealing Directive 2000/55/EC of the European Parliament and of the Council, CEC, Brussels, 2009. Available at http://eur-lex.europa.eu/LexUriServ/LexUriServ.do?uri=OJ:L:2009:076:0017:0044:EN:PDF.

18 Wilkins A J, Nimmo-Smith I, Slater A I and Bedocs L. Fluorescent lighting, headaches and eyestrain. Lighting Research & Technology, 1989, 21 (1) 11–18.

19 Tesco's need for speed. Lux Magazine, March 2012, 20–24.

20 Carbon Trust (CT). Display lighting guide CTG010. London, The Carbon Trust, 2012.

21 California Energy Commission (CEC). Daylight and retail sales. Technical report. Sacramento, CA, CEC, 2003.

22 Littlefair P J. Designing with innovative daylighting. Bracknell, IHS BRE Press, 1996.

23 Monodraught. The importance of natural ventilation and daylight for retail applications. High Wycombe, Monodraught, 2010.

24 Littlefair P J. Selecting lighting controls. BRE DG 498. Bracknell, IHS BRE Press, 2006.

25 Zumtobel studies store lighting and shoppers' behaviour. Zumtobel, 2012. Available online at http://lighting.com/zumtobel-store-lighting/?goback=%2Egde_3834245_member_92857325.

26 Faithful M. This isn't any store... Lux Magazine, May 2011, 28–31.

27 Faithfull M. Good retail lighting does add to the bottom line – official. Lux Magazine, June 2011, 47.

28 Philips Lighting. Case study: PLUS Heerlerheide. Amsterdam, Philips Lighting, 2010. Available online at www.lighting.philips.com/pwc_li/main/shared/assets/downloads/CaseStudy-Plus-Heerlerheide-INT.pdf.

29 Burgess M. Low-energy luxury. Lux Magazine, May 2011, 60–61.

30 Turner J. Designing with light: Retail spaces. Crans-Près-Céligny, Switzerland, RotoVision, 1998.

31 Baumstarck A. Effects of dressing room lighting direction on consumers' perceptions of self and environment. Master of Interior Design Thesis. Gainesville, FL, University of Florida, 2008.

32 Entwistle J. Boux's selector. Lux Magazine, June 2011, 48.

33 BSI. Lighting of work places. Outdoor work places. BS EN 12464-2:2007. London, BSI, 2007.

© iGuzzini: Repower, Ilanz, Switzerland

Appendix: Checklist for refurbishment

Area	What to look for	What to do
General retail areas	Lighting more than 15 years old. Fluorescent tubes with bayonet cap connections. Badly discoloured plastic diffusers.	High priority for replacement. Get an expert lighting survey carried out.
	38 mm (1.5 in) diameter T12 fluorescent tubes in switch-start fittings.	Fit 26 mm (1 in) diameter T8 fluorescent tubes as direct replacements in open fittings, or replace enclosed fittings with new fittings using 16 mm (5/8 in) diameter T5 lamps.
	26 mm (1 in) diameter T8 fluorescent tubes in enclosed fittings.	Consider replacing with new T5 fluorescent fittings.
	600 mm^2 fittings with 55 W CFLs.	Replace with modern reflector or high-performance prismatic lens fittings using T5 triphosphor fluorescent lamps or LED fittings.
	PAR 38 sealed-beam reflector lamps. Tungsten halogen lighting.	Replace with metal halide discharge lighting or LED lighting.
'Warehouse' type retail premises	High-pressure mercury lamps. Old 8 ft 125 W fluorescent fittings.	For most high-bay applications, replace with metal halide lighting. For low mounting heights, replace with modern T5 fluorescent or LED lighting.
Lighting controls	Large areas controlled by a switch near the door. Lighting on outside normal working hours. Potential for increased use of daylighting.	Install appropriate lighting controls to enable lighting to be switched off when not needed, or dimmed in daylit areas.
External lighting	Tungsten halogen lamps. Signboard, security or car park lighting on during hours of daylight.	Install new lighting using metal halide or LED lamps. Install automatic time and/or daylight-linking controls.
Floodlighting	High-wattage tungsten halogen floodlights.	Replace with metal halide or LED lighting.
Staffrooms, stockrooms and utility rooms	Check lights not left on when unoccupied.	Post reminder notices near switches and on noticeboards. Appoint member of staff with responsibility for checking. Consider installation of automatic lighting controls using time switches and/or occupancy sensors to vary light outputs according to use.
Good housekeeping	Are lamps and luminaires clean and unobstructed? Are the staff aware of the need for energy efficiency?	Ensure maintenance schedule includes regular lamp and fitting cleaning routines. Instigate information and energy awareness campaign among staff.

Other reports from BRE Trust

Modern methods of house construction: a surveyor's guide. **FB 11**

Crime opportunity profiling of streets (COPS): a quick crime analysis– rapid implementation approach. **FB 12**

Subsidence damage to domestic buildings: a guide to good technical practice. **FB 13**

Sustainable refurbishment of Victorian housing: guidance, assessment method and case studies. **FB 14**

Putting a price on sustainable schools. **FB 15**

Knock it down or do it up? **FB 16**

Micro-wind turbines in urban environments: an assessment. **FB 17**

Siting micro-wind turbines on house roofs. **FB 18**

Automatic fire sprinkler systems: a guide to good practice. **FB 19**

Complying with the Code for Sustainable Homes: lessons learnt on the BRE Innovation Park. **FB 20**

The move to low-carbon design: are designers taking the needs of building users into account? **FB 21**

Building-mounted micro-wind turbines on high-rise and commercial buildings. **FB 22**

The real cost of poor housing. **FB 23**

A guide to the Simplified Building Energy Model (SBEM): what it does and how it works. **FB 24**

Vacant dwellings in England: the challenges and costs of bringing them back into use. **FB 25**

Energy efficiency in new and existing buildings: comparative costs and CO_2 savings. **FB 26**

Health and productivity benefits of sustainable schools: a review. **FB 27**

Integrating BREEAM throughout the design process: a guide to achieving higher BREEAM and Code for Sustainable Homes ratings through incorporation with the RIBA Outline Plan of Work and other procurement routes. **FB 28**

Design fires for use in fire safety engineering. **FB 29**

Ventilation for healthy buildings: reducing the impact of urban pollution. **FB 30**

Financing UK carbon reduction projects. **FB 31**

The cost of poor housing in Wales. **FB 32**

Dynamic comfort criteria for structures: a review of UK standards, codes and advisory documents. **FB 33**

Water mist fire protection in offices: experimental testing and development of a test protocol. **FB 34**

Airtightness in commercial and public buildings. 3rd edn. **FB 35**

Biomass energy. **FB 36**

Environmental impact of insulation. **FB 37**

Environmental impact of vertical cladding. **FB 38**

Environmental impact of floor finishes: incorporating The Green Guide ratings for floor finishes. **FB 39**

LED lighting. **FB 40**

Radon in the workplace. 2nd edn. **FB 41**

U-value conventions in practice. **FB 42**

Lessons learned from community-based microgeneration projects: the impact of renewable energy capital grant schemes. **FB 43**

Energy management in the built environment: a review of best practice. **FB 44**

The cost of poor housing in Northern Ireland. **FB 45**

Ninety years of housing, 1921–2011. **FB 46**

BREEAM and the Code for Sustainable Homes on the London 2012 Olympic Park. **FB 47**

Saving money, resources and carbon through SMARTWaste. **FB 48**

Concrete usage in the London 2012 Olympic Park and the Olympic and Paralympic Village and its embodied carbon content. **FB 49**

A guide to the use of urban timber. **FB 50**

Low flow water fittings: will people accept them? **FB 51**

Evacuating vulnerable and dependent people from buildings in an emergency. **FB 52**

Refurbishing stairs in dwellings to reduce the risk of falls and injuries. **FB 53**

Dealing with difficult demolition wastes: a guide. **FB 54**

Security glazing: is it all that it's cracked up to be? **FB 55**

The essential guide to retail lighting. **FB 56**

Environmental impact of metals. **FB 57**

Environmental impact of brick, stone and concrete. **FB 58**